The complexity of strains and displacements at plate boundaries results in fragments of the Earth's crust – ranging from tens of metres to tens of kilometres in size – travelling up to thousands of kilometres with respect to each other. These terranes are not subducted but are accreted to the margin of the continent under which subduction is taking place. Orogenic belts are viewed as consisting of many terranes, including slivers of oceanic crust and older continental basement. Evidence for the displacement of the terranes comes from palaeomagnetic and faunal palaeolatitudinal data.

In November 1989 the Royal Society Discussion Meeting on Allochthonous Terranes was convened to assess the past, present and future significance of the terrane concept in understanding the processes at work at plate boundaries. Contributions discuss exotic terranes around the world throughout the geological timescale.

Allochthonous terranes

ALLOCHTHONOUS TERRANES

EDITED BY

J. F. DEWEY
Department of Earth Sciences, University of Oxford

I. G. GASS
Department of Earth Sciences, The Open University, Milton Keynes

G. B. CURRY
Department of Geology, University of Glasgow

N. B. W. HARRIS
Department of Earth Sciences, The Open University, Milton Keynes

A. M. C. SENGÖR
ITU, Istanbul, Turkey

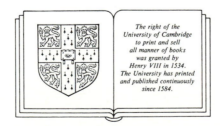

The right of the
University of Cambridge
to print and sell
all manner of books
was granted by
Henry VIII in 1534.
The University has printed
and published continuously
since 1584.

CAMBRIDGE UNIVERSITY PRESS
Cambridge
New York   Port Chester
Melbourne   Sydney

Published by the Press Syndicate of the University of Cambridge
The Pitt Building, Trumpington Street, Cambridge CB2 1RP
40 West 20th Street, New York, NY 10011-4211, USA
10 Stamford Road, Oakleigh, Melbourne 3166, Australia

First published by the Royal Society 1990 as Volume 331 Number 1620
of *Philosophical Transactions of the Royal Society*, A,
and © 1990 The Royal Society and authors of individual papers

First published as a book by Cambridge University Press 1991

© Cambridge University Press 1991

Printed in Great Britain at the University Press, Cambridge

British Library cataloguing in publication data

Allochthonous terranes.
1. Tectonics
I. Dewey, J. F. (John Frederick)
551.136

Library of Congress cataloguing in publication data

Allochthonous terranes/edited by J. F. Dewey … [et al.].
p. cm.
'First published in 1990 as volume 331, number 1620 of
Philosophical transactions of the Royal Society' —CIP galley.
Includes index.
ISBN 0-521-40461-4
1. Thrust faults (Geology) I. Dewey, J. F. (John F.)
QE606.A46 1991
551.8′7—dc20 91-2542 CIP

ISBN 0 521 40461 4 hardback

CONTENTS

CONTENTS

PREFACE

The terrane concept and its practice, terranology, were developed by geologists, principally P. Coney, D. L. Jones and J. W. Monger, in assessing the palaeogeographic coherence of the North American Cordillera System from Mexico to Alaska. The notion that substantial chunks of Alaska and the Canadian Cordillera are 'exotic' with respect to the North American Craton and might have travelled northwards for thousands of kilometres was formulated first from palaeomagnetic and faunal palaeolatitudinal data. This was supported later by palaeomagnetic evidence of large and rapid clockwise Cenozoic rotations of large areas on the oceanward 'outboard' margin of the Cordillera. The terranologists incorporated, synthesized and analysed a huge amount of stratigraphic and faunal (most importantly radiolarian) data to show that much of the Cordillera, especially, but not exclusively, the outboard western margin, is composed of blocks and slivers (terranes) ranging from tens of metres to tens of kilometres and that, although palaeotectonic history can be elucidated for individual terranes, their original positions relative to one another and to the North American Craton have remained difficult to determine. Many terranes are clearly of oceanic origin consisting of island arcs, sea mounts and sea mount chains swept in from the Pacific whereas others have older continental basements and have travelled northwards along the continental margin much as the long narrow Salinian block (terrane) is moving northwards today along the San Andreas Fault.

The idea that orogenic belts are made up of displaced fragments is inherent of course in plate tectonics. If plates have moved thousands of kilometres with respect to one another and the relative motion is not taken up along a single line during the whole of the motion, there is bound to be substantial motion of slivers within the plate boundary zone relative to one or both plates. Another way of putting this is that plate displacement is converted and partitioned into plate boundary strain and displacement in complicated ways that may change rapidly with time as a result of triple junction migration, rotational and 'locking' compatibility problems, strain hardening and rheological changes resulting from cooling and heating. Atwater's classic inceptive (1970) paper on the Mesozoic/Cenozoic interaction of the Pacific oceanic plates with western North America was the earliest harbinger of the terrane concept in developing a plate tectonic quantitative logical framework for Cordilleran evolution that demanded and demonstrated terrane mobility in a pattern of rapid changes from convergence to strike-slip dischronously along the Cordilleran trend as triple junctions evolved. Atwater's paper also occupies another fundamental position in the history of geological ideas in being the first publication to link, quantitatively and convincingly, relative plate motion with continental geology.

Inherent in the terrane concept is that coherent plate tectonic patterns of, for example, arcs, forearcs and trenches, are subsequently or even synchronously disrupted into a mosaic of smaller fragments (terranes). The terranes are then reassembled, commonly by very large strike-slip motions so that excision and repetition of earlier more coherent plate tectonic patterns (commonly called Dewey-grams by the more ardent terranologists) is accompanied by rotation and deformation of the terranes. Thus, transverse orogenic palaeogeographic sections that do not restore the terranes to their original relative positions are meaningless. In this logic, the terrane is a coherent tectonostratigraphic geographical unit bounded by structural contacts across which displacement matching cannot be achieved. Terrane boundaries like the Highland Boundary Fault Zone in Scotland commonly contain, therefore,

PREFACE

a remarkably diverse range of exotic lithologies that represent fragments excised from the various regions past which the terrane moved.

In terrane philosophy, coherent tracts of an orogenic belt are regarded as being suspect until 'innocent' autochthoneity with respect to an adjacent terrane or craton can be established. If innocence cannot be proved or can be disproved, the terrane becomes guilty and the final result is that the orogenic belt becomes a jumble of terranes in which the search is then on for the degree of allochthoneity (displaced, far travelled, exotic), which must, by definition, be always greater than the length of the bounding faults. With terranology has sprung up the new terminology of detaching, travelling, docking, stitching and linking to emphasize and describe terrane mobility.

Terranology has had a range of responses from sycophantic adherence on the one hand to scornful dismissal on the other, with various intermediate views such as robust practical working use, tentative, or faintly sceptical. This discussion meeting on Allochthonous Terranes was convened to assess, ten years on, the past, present and future significance and role of the terrane concept and terranology in understanding plate boundary processes, particularly the evolution of orogenic belts. Particular emphasis has been placed in this meeting on the tectonic mechanisms that lead to the generation of terranes and to analysing the geological history of a wide range of regions to evaluate the possible role of terranes in their evolution and the usefulness of terrane methodology in analysing and synthesizing their tectonics.

John F. Dewey

Terranology: vice or virtue?

By A. M. C. Şengör[1] and J. F. Dewey[2], F.R.S.

[1] *I.T.Ü. Maden Fakültesi, Jeoloji Bölümü, Ayazaga* **80626**, *Istanbul, Turkey*
[2] *Department of Earth Sciences, Oxford University, Oxford OX1 3PR, U.K.*

The concept of terranes and the methodology of terranology are not intrinsically new ideas and methods that have advanced our understanding of orogenic and plate boundary processes. They are, rather, a new formulation of ideas inherent in earlier concepts of displacement partitioning in plate boundary zones and much earlier ideas about orogenic elements and zones. Terranology has had the useful effect of emphasizing, and awakening the tectonic community at large to, the possibility of very large orogen parallel strike–slip motions that predate continental collisions.

Introduction

The recognition, by Wilson (1966, 1968), Moores (1970), Monger & Ross (1971), Dewey *et al.* (1973), and Roeder (1973), among others, of the interlacing mosaic-like geometry and the complex evolution of orogenic belts (see, for example, Dewey 1975, 1976), was reformulated recently in the form of the mainly descriptive allochthonous or suspect tectonostratigraphic terrane concept (Monger 1975; Coney 1978, 1981, 1989; Coney *et al.* 1980; Kerr 1980; Ben-Avraham *et al.* 1981; Jones *et al.* 1982, 1983 *a, b*; Saleeby 1983; Howell 1989; Silbering & Jones 1984). This idea holds that major orogenic belts consist of 'collages' (Helwig 1974) of fault-bounded crustal and/or lithospheric fragments, termed terranes, of diverse origin and various sizes, the continuous movements of which, including dispersal and reunion through sundry tectonic processes, are believed to have imparted on the orogenic belt a mobility supposedly hitherto unrecognized and largely unsuspected. For this 'late' recognition, 'classical' (Coney 1981, p. 27) plate tectonic models have been blamed, which are considered to be 'either too facile or inappropriate when applied to complex geologic settings for two reasons: (1) they are two dimensional and ignore the complexities inherent in systems rapidly evolving in space and time; and (2) they assume that genetic relations must exist between adjoining but different domains (terranes). This assumption has led inexorably to confusion between fact and interpretation...' (Jones *et al.* 1983 *a*, p. 103). These geologists were led to the conviction that the plate tectonic approach to orogenic belts had been dominated by models that were oversimplified, deterministic, if not entirely naive, and that a new methodology, involving objective criteria to recognize individual terranes and to 'prove' their spatial relationships before any plate tectonic interpretation could be attempted, was necessary (Schermer *et al.* 1984, p. 112). This 'new' methodology has been called terrane analysis (Jones *et al.* 1983 *a, b*; Schermer *et al.* 1984; Howell 1985 *a*, 1989; Howell *et al.* 1985) and has spread to other orogenic belts applied both by the originators of the concept (see, for example, Ben-Avraham *et al.* 1981; Nur & Ben-Avraham 1982 *a, b*; Nur 1983; Schermer *et al.* 1984; Howell 1985 *a*, 1989; Howell *et al.* 1985; Xiang & Coney 1985) and by others, for example, in the Appalachians (Williams & Hatcher 1982, 1983), the British Caledonides (Hutton & Dewey

1

1986; Bentley *et al.* 1988; Mason 1988; McKerrow & Gibbons 1988); the Scandinavian Caledonides (Stephens & Gee 1985); the European Hercynides (Weber 1986); the Tethysides of the Tibetan–Himalayan region (Chang *et al.* 1986); the Qin-Lang (Jia *et al.* 1988); Chinese orogenic belts in general (Guo *et al.* 1984); Yenshanian orogeny (Ye *et al.* 1984); the Alps (Tollmann 1987); Proterozoic orogeny (Grambling *et al.* 1988); and Circum-Pacific orogenic belts (Hashimoto & Uyeda 1983; Howell 1985*b*; Leitch & Scheiber 1987; Monger & Francheateau 1987). Because, in the past few years, interpretation of continental tectonics in terms of terranes has been such a dominant topic, reactions to which range from enthusiastic applause to abusive rejection, it is useful to review 'terranology' in the light of the following questions: (1) Is the terrane a new concept or jargon for old methods and ideas? (2) Is it a better word than fragment, block, sliver, etc.? (3) Has it been dangerous or obscuring in any way and/or has it had useful effects? Is terranology a useful methodology? (4) Does it have some objective methodology not inherent in other 'methods' of field work, regional geology, data analysis and synthesis? (5) Is it model-independent, and, if so, is this useful? (6) Is it a new way of looking at orogens that gives new insights and does it, as maintained by some (e.g. Howell 1985*c*), go beyond plate tectonics? (7) Does it advance our understanding of processes in tectonics? (8) Are terrane and terrane analysis temporary or permanent concepts? To attempt answers to these questions we summarize the evolution of thought (Sengör 1990*a*) on three dimensional mobility in orogenic belts, which allegedly culminated in the theory of terranes, to show that the stage of development represented by terranology had been reached and surpassed in at least one orogenic belt even before the advent of plate tectonics. We then look at our present understanding of orogenic mobility to evaluate the role of terranology.

EVOLUTION OF THOUGHT ON OROGENIC MOBILITY

That mountain belts consist of numerous fault-bounded blocks was the first general conclusion reached by geologists, especially those working in the Alps. Although this view accounted for local distribution of rock types and structures by postulating *ad hoc* fault-bounded blocks as need arose, it made delineation and predictions of a general orogenic architecture exceedingly difficult. The faults were ascribed to vertical uplift, to which the entire orogenic mobility was reduced.

Arnold Escher von der Linth's observations in the Swiss Alps in the first half of the 19th century showed that much of the great variability in Alpine geology was more a result of complicated folding than faulting. Escher's postulate of a regularly arranged set of folds being responsible for the main outlines of Alpine structure also was in agreement with the somewhat later results of other workers' studies in other mountain ranges (see, for example, Rogers & Rogers 1843) in the Appalachians, de la Beche (1849) in Southwest England and Wales). Mainly through the work of Swiss and American geologists, the view that mountain belts are compressed and folded portions of the Earth's crust became the conventional wisdom, to the point of calling all major mountain chains 'foldbelts'. Following Elie de Beaumont's analogy, the formative mechanism was likened to the operation of the jaws of a vice, introducing a second, horizontal, dimension into orogenic mobility. The ideal foldbelt was held to be symmetric with respect to an axis following its crest.

Suess (1875) was the first to argue convincingly that mountain belts had an asymmetric structure verging mainly towards their 'external' sides, towards which most orogens were convex. This idea of asymmetry in orogenic architecture was difficult to reconcile with postulated large, symmetric folds, of which the Glarus double-fold had become perhaps the most famous. In 1884 Marcel Bertrand reinterpreted the Glarus double-fold, in the light of Suess's views, as a single north-vergent nappe indicating a stratal shortening of some 40 km. With Bertrand's nappe hypothesis, the two-dimensional (uplift and subsidence with shortening across the trend of orogens) evolution of orogens became entrenched, to which Suess (1891) later added an extensional component on the basis of his study of the East African Rift Valley.

In the light of Bertrand's idea, many hitherto-unexplained structural and stratigraphic anomalies of Alpine tectonics became intelligible (Sengör 1990b). In particular, Schardt (1893) used the nappe idea to solve the riddle of the *Préalpes*, where rock assemblages had been mapped, that appeared completely exotic with respect to their surroundings. The then 'orthodox' hypotheses to explain their origin ranged from barely plausible sunken mountain ranges (see, for example, Studer 1843) to the entirely fantastic transport of whole mountain massifs by now-extinct, monstrous rivers (see, for example, Früh 1888). Schardt (1893) showed that, if these totally 'foreign terranes' containing coeval but utterly different rock types and structures with respect to the rocks amidst which they were now located, were viewed as remnants of a large nappe of southerly provenance, the riddle of their emplacement could be solved without resorting to improbable means. Schardt's views were adopted and amplified by Lugeon (1902), who showed that the entire northern margin of the Alps represents a stack of nappes of southerly provenance and speculated that the internal, metamorphic parts of the chain also consisted of north-vergent, gneiss-cored nappes. Lugeon argued that, by 'undoing' the deformation represented by the nappe pile, one could reconstruct the pre-orogenic palaeogeography of the Alps. To guide such a reconstruction, Lugeon (1902) devised the 'rule' that 'the higher a nappe is located in the pile and the farther it now extends northwards, the farther to the south it must root'. Application of Lugeon's rule necessitated the completion of the mapping of all the major nappes in at least one cross section across the Alps. Argand completed this task, which culminated in the discovery of the highest nappe in western Switzerland. To reproduce the pre-orogenic palaeogeography by undoing the Alpine deformation, Argand combined Lugeon's kinematic rule with a stratigraphic rule inspired by Haug (1907). Argand (1911, 1916) maintained that every major nappe had a uniquely stratigraphy reflecting a geological history different from its neighbouring nappes. The basic pattern of all the major Penninic nappes was characterized, according to Argand (1916), by neritic sedimentary rocks occupying the nose region of the nappe, 'comprehensive' geosynclinal sequences in the back and foredeep clastics with local breccias under the nappe. Applying Lugeon's rule to this general stratigraphic picture, Argand concluded that it reflected the order of original distribution of tectono–sedimentary environments in the 'Tethys', which was caused by the northerly march of giant recumbent folds, the 'nappe embryos', that became exaggerated later into the present Pennine nappes (Argand 1916, fig. 1). Argand's theory was an immediate success and remained the only working hypothesis well into the 1940s, despite some dissension from outside Switzerland (mainly by Haug 1925).

As a result of his 'embryotectonics', postulating a continuous shortening of the Alpine Tethys between the Carboniferous and the present, Argand quickly acknowledged that the prevailing 'fixist' theories of orogeny could not accommodate his scheme and he opted for continental drift (Argand 1916, 1924). The highly sinuous outline of the Alpine orogenic belt led him to the hypothesis of colliding irregular continental margins, a consequence of which was strike–slip movement subparallel with the overall trend of the orogenic belt (which Argand called 'flow along the shore' alluding to the margin of the 'Alpine geosyncline' (Argand 1924, fig. 14).

Argand's theory finally brings us to a three-dimensional theory of orogenesis, in which movements both across (in vertical and in horizontal planes) and in horizontal planes) and along the orogens were acknowledged. With the advent of the idea of nappes and its elaboration by Argand's application of his embryotectonics to continental drift, Alpine tectonics reverted back to its starting point, in being viewed as a pile of fault-bounded blocks. Instead of the dominantly steep to vertical faults of the late 18th and early 19th century, however, faults separating the nappes were thought to be dominantly roughly horizontal. After Argand, the standard procedure in alpine tectonic studies became the recognition of fault-bounded blocks with geological histories distinct from surrounding blocks. Contrary to conventional wisdom, all three kinds of faults were believed to surround the nappes, and displacements exceeding tens of kilometres on each type had been mapped on each type (Sengör 1990b). Argand (1924, figs 26 and 27) also recognized post-orogenic disruption because of ongoing continental drift and the complexities implicit in its kinematics, although most Alpine geologists did not follow his lead until the rise of plate tectonics mobilism.

The history of thought in Alpine tectonics up to this point makes it clear that the answer to the first question posed above, whether the 'terrane' is a new concept or new jargon for old methods and ideas, has to be the latter. The answer to whether it is a useful word *versus* nappe, strike–slip duplex, fragment, block, sliver, etc., is generally 'no'. If a fault-bounded package can be identified for what it is, why not call it that? Structural geology has developed a rich nomenclature for fault-bounded blocks, depending on the nature, attitude, and displacement of their bounding faults and it would be a retrogressive step to employ a lump term for them all. The word 'terrane' is an unfortunate choice to designate tectonic entities because it has long been used in the geological literature as an informal *stratigraphic* designation (also as *terrain*, but both the **1989** edition of the *Oxford English Dictionary* and *Webster's Third New International Dictionary of the English Language*, **1966**, prefer the spelling terrane when the term is used in geology).

Before biostratigraphy, in the late 18th and early 19th century, terrane referred to the ensemble of all formations dominated by the same rock type implying coeval deposition in the Neptunist framework. Later, this definition changed following a change in providing views about Neptunism, especially of its temporal connotation. In 1828, D'Aubuisson de Voisins (pp. 268–269) wrote: 'The word *terrane* is often taken as a synonym of *formation*. However, its employment is more multifarious and less precise, especially regarding the epoch of production....One could say that in geognosy formations are the species and, up to a point, terranes are the genera.' In 1841, Dufrénay & de Beaumont (1841, p. 35) stated that '...beds, related to one another by a regular succession which is remarkable and formed under the same conditions constitute what is called a terrane...The expression terrane therefore always implies an ensemble as opposed to *rock*, which by contrast, refers to the homogeneous material of each bed taken in isolation'. In the continental literature, the term terrane gradually came to signify a *stratigraphic system*, but the *First International Geological Congress* recommended that system be preferred. Usage in English always has been similar to D'Aubuisson de Voisins's 1828 definition quoted above. For example, Dana (1880, p. 81) wrote: '*terrane*...is used for any single rock or continuous series of rocks, of a region, whether the formation be stratified or not.

It is applied especially to metamorphic and igneous rocks, as a *basaltic terrane*, etc....'. Since the last century, the term terrane has now been used in geology in this imprecise sense. In the 1989 edition of the *Oxford English Dictionary*, it is defined as 'a name for a connected series, group, or system of rocks of formations, a stratigraphic subdivision'. *Webster's* (1966) gives a similar definition: 'a rock formation or a group of formations...' the area or surface over which a particular rock or group of rocks is prevalent'.

The AGI *Glossary of geology* (Gary *et al.* 1972) defines terrane as an obsolescent term applied to a rock or group of rocks and to the area in which it outcrops. The term is used in a general sense and does not necessarily imply a specific rock unit or group of rock units.' However, a perusal of the literature of the early 1970s shows that the term was not at all obsolescent! The future promulgators of tectonic terranes themselves used terrane in its old stratigraphic sense as late as 1972 (see, for example, Jones *et al.* 1972, caption to fig. 3).

Thus the English and French word *terrane* has been used both geographically and stratigraphically. Because these older usages of terrane in geology remain current, there is no need to add a vague tectonic meaning of the word terrane to the already vague geographic and stratigraphic meanings. It is our opinion that the words block (that conjures up the image of a roughly equant rather angular shape), sliver (an elongate object with pointed ends), or fragment (implying an object broken off a larger piece) are more informative than terrane, because the latter simply denotes a nondescript surface or expanse, and the former are not burdened by any previous geographic or stratigraphic connotation. We suggest that the genetic terms nappe, extensional allochthon, strike–slip sliver, fragment, and the non-genetic but geometrically informative terms sliver and block are preferable to terrane.

There is another category of genetic terms that may be used in place of terrane. To consider those, we need to go back to the history of thought on orogenic mobility and pick up the story from the time when Argand's views were refuted. The main breakthrough in Alpine tectonics came with the criticism and eventual refutation of Argand's views. In the 50s, it was shown that no one-to-one correspondence exists between the 'pre-orogenic' palaeogeographic entities, originally interpreted by Argand as nappe embryos (which he compared with present-day island arcs: Argand 1924) and the later nappes (Trümpy 1955, 1960). What created the former had been extension and not shortening as Argand has supposed; the nappes formed under a later, compressional régime. Trümpy (1960) summarized this revolution in Alpine thinking in terms of the recognition that the 'pre-orogenic' structures, and the strain régime that created them, had had no relationship with the 'orogenic' structures and the associated strain régime. To interpret the geological history of the Alps correctly, Trümpy argued, one had to distinguish a 'pre-orogenic' group of objects and an 'orogenic' one, whose relationships were complex.

After the advent of the theory of plate tectonics, it became clear with what these two groups of objects correspond; the former group consists of microcontinents, island arcs, ocean floors, and continental margins, whereas the latter was formed from the fault-bounded rock packages generated by the deformation of a collage consisting of the members of the former group. In a plate tectonic interpretation of the Alpine System, Dewey *et al.* (1973) recognized, and made extensive use of, this distinction. Their various microcontinents, island arcs, or small ocean basins, for example, do not correspond in any way with the individual nappes that Argand had established (Dewey *et al.* 1973, figs 1 and 12). In fact, fragments of Dewey *et al.*'s (1973) *primary orogenic collage components* are now found distributed in a number of *secondary orogenic collage components* that include the major Alpine nappes (cf. Sengör 1990 *a, b*). This same distinction was made by Wilson (1968) in the western North American Dordillera, where he distinguished offshore island arcs and a western land that impinged on the original margin of Western North America and distinguished those from 'fragments' that formed by 'great dislocation between the Cordilleras of Canada and those of the United States' (Wilson 1968, p. 316).

With mobilism, especially after the advent of plate tectonics, fault-bounded entities making up an orogenic belt were seen to be of two fundamentally different kinds. One consists of non-subductable objects, such as island arcs, microcontinents, or large accretionary complexes that collide with and become accreted to a continental nucleus and/or to each other, forming the primary orogenic components. Before, during, and after such accretion the primary orogenic collage components become chopped into nappes, strike–slip duplexes, and extensional allochthons, individually or collectively, forming the secondary orogenic collage components. The faults that accomplish this chopping may have considerable displacements on them, although both the normal and thrust varieties have dynamic limits to their displacements; only strike–slip faults have no such dynamic limits. In studying the evolution of orogenic mosaics, our most important goal therefore becomes the identification of its primary collage components first, as precisely as we can. We then use them to study the effects of the secondary disruption, to which the primary components are related as strain markers.

Figure 1 shows schematically and in a very simplified sequence the formation of primary and secondary orogenic collage components. Terranology is not only not new, but it represents a return to a tried and abandoned methodology. Some may find this perfectly permissible in an anarchic philosophy of science, in which 'anything goes' (Feyerabend 1988), but it is our view that already-discarded methods of inquiry seldom bear fruit when rehabilitated. Terranology

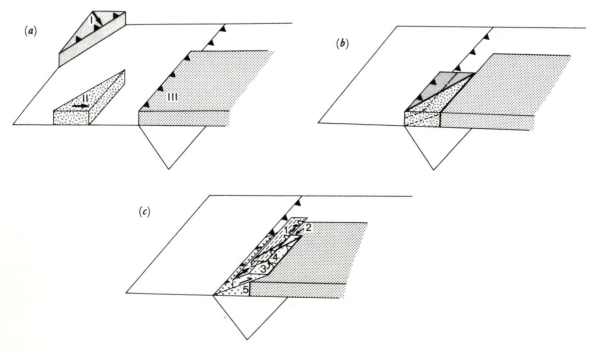

FIGURE 1. Sketch illustrating the relationship between primary (I, II, III) and secondary (1, 2, 3, 4, 5) collage components.

not only does not go beyond plate tectonics, it takes a backward step; by confusing primary and secondary collage components, it confuses also their genetic implications.

In the following section, we outline the plate tectonic causes of orogenic collage formation and argue that different sorts of collages form and evolve along different sorts of orogenic belts that are distinguished from one another by (1) the size of ocean lost along them, (2) the number of plates involved, and (3) the dominant orientation of convergence vector along the orogen during its history.

COMPLEXITIES OF OROGENIC EVOLUTION

(a) Complexities generated by finite plate evolution

A persistent terranological criticism of 'classical' plate tectonic methodology was that it had remained largely two dimensional and had been, therefore, unable to account for complexities inherent in systems rapidly evolving in space and time. Atwater (1970) was the first to emphasize the tremendous complexities that arise as a consequence of the finite evolution of three plates, which inevitably results in an extremely complicated history recording complex overprinting relationships among rapidly changing tectonic regimes. Complications also arise in a very simple 2-plate system that evolves without involving any change in relative motion rate or direction (Dewey 1976, figs 1 and 2). The stratigraphic correlation chart (Dewey 1976, fig. 2) reveals a history much more complicated than is intuitively expected from such a simple system. Localities 45 and 55, for example, become neighbours on the same plate although they have drastically different geological histories, and are now separated by a fossil fault. The same is true for localities 31 and 58, and 10 and 60. The faults separating localities 10, 31, and 45 from 55, 58 and 60 juxtapose points created far away from one another, thus satisfying 'the important criterion...that movement be sufficient to completely disrupt original facies relations and thus render uncertain original genetic relations' (Schermer et al. 1984), p. 118). These localities, especially 10, 31, and 45, must be separated from each other by faults also because of the peculiarities of fracture zone evolution (De Long et al. 1979). Along a single, steady, and stable plate boundary, relative plate motion of the simplest type thus can create 'fault-bounded geologic entities...with geological histories different from neighbouring localities'.

If three or more plates are in relative motion, at least one boundary must shift with respect to the instantaneous pole of rotation between neighbouring plates implying that at least one boundary must change character continuously (Dewey 1975). This change in character may in places lead to complex boundary configurations between adjacent plates, especially if non-subductable material is carried by at least one of them (Dewey 1975, fig. 14). In extreme cases, plates break up and form smaller plates (such as happened to the Caribbean plate (Pindell & Dewey 1982)). Pole shifts require that transform faults also change character or shift. If such shifts occur as jumps, exchanges of areas between plates may occur (Dewey 1975).

The discussion so far has concentrated on showing how complicated orogenic evolution must be if it results from finite plate evolution. But all these complexities are still at the plate level only and are not relevant to intraplate or inter-plate boundary zone phenomena. In the next section, we discuss briefly the consequences of superimposing on these plate-level complexities, some geological phenomena that dominantly result from the changing rheological properties of the lithosphere both horizontally and vertically in response to changes in rock types, pre-existing structural fabric, temperature, pressure, and availability of water.

(b) Complexities generated by geological processes

In 1975, Dewey observed that 'It is a truism that orogenic belts are exceedingly complex and poorly understood. If orogenic belts are the result of the relative motion of plate mosaics carrying continents and island arcs, this is not surprising. The writer has heard, on numerous occasions, the criticism that the application of plate tectonics in attempting to analyse orogenic belts grossly oversimplifies the geology of orogenic belts. In the writer's view this emphasizes the scale problem. It is one thing to say that a blueschist-melange marks the site of a former subduction zone but quite another to relate, in detail, the structural and metamorphic evolution of that zone to changing rates and directions of subduction. The broad geologic corollaries of plate tectonics are a matter of observation; the fine subtle corollaries that should now concern us are not understood. It is the main purpose of this paper to show that the evolution of plate mosaics carrying continents and island arcs must generate deformation sequences so complex that, with the difficulties of establishing relative plate motion solutions for the past, they may never be completely understood in plate tectonic terms' (Dewey 1975, p. 262). It is impossible to review, here, all the 'subtle' geologic corollaries of plate tectonics that may create a variety of 'terranes', but one can summarize them under four headings as follows.

(i) *During rifting.* Rifting is a complex process that takes place at all stages of a Wilson cycle. It not only creates volcanic and non-volcanic Atlantic-type continental margins (White *et al.* 1989), it also generates microcontinents of diverse shapes and aspect ratios, that may be generated in company of abundant or sparse magmatism, and have different subsidence histories. All these microcontinental pieces are eventually caught up in orogeny, forming various classes of terranes. Extant examples include the Lomonosov Ridge, Madagascar, the Seychelles, the Nazarene and the Saya de Malha blocks, the Danakil Horst, Sardinia–Corsica block, Lord Howe Rise, Norfolk Ridge, and Yamato Bank. In orogenic belts, such microcontinents are well known from the Tethysides, ranging in size from the immense Cimmerian Continent occupying a considerable length of the orogen to the tiny Brianconnais and Margna fragments in the Alps.

(ii) *During strike–slip faulting.* Strike–slip faulting is a process that also occurs at all scales and during all stages in a Wilson cycle. It commonly creates slivers during the establishment of a through-going strike–slip fault by linking up Riedel and P shears and then breaking new faults across the originally sinuous trace, or isolating strike–slip duplexes by reactivating older parallel and sub-parallel fault segments that lie along its course. Crack propagation may give rise to rhomb-shaped strike–slip duplexes while the main throughgoing fault is forming (Bhat 1983). Stick–slip behaviour and its inhomogeneous distribution along an established strike–slip fault zone may throw the fault off course, generate a sinuous trend, and the breaking of a new straight segment across the sinuous trace may result in new strike–slip duplexes.

Both positive and negative flower structures form along strike–slip faults depending on the regional and local orientation of the slip vector with respect to the fault trace. Positive flowers tend to generate flakes that are progressively detached from their roots and rotate, by up to 180°. In the course of their rotation, they may disintegrate and form a number of small flakes and klippen that come to rest on totally foreign substrates. By contrast, negative flowers tend to form blocks that rotate like ball-bearings or like circular disks rotating, where no block coupling exists, with an angular velocity of half of the vertical component of the vorticity of an underlying fluid (McKenzie & Jackson 1983). If the shape of the generated block in a negative flower is more like an ellipse than a circle, the rotation will be much slower than just half the vorticity and will depend on the aspect ratio of the block (Lamb 1987). In an evolving shear zone, Riedel–Anti-Riedel pairs may isolate blocks or slivers that rotate with the sense of shear.

If various rift segments are linked by long transform fault segments during continental separation, imperfections along the transform traces may result in the ripping of rectangular or triangular blocks from the separating corners of the continental margins (Scrutton 1976).

If strike–slip motion occurs parallel with a pre-existing continental margin, the fault that localizes the motion will be located inland of the ocean–continent interface. Its precise location is a function of the original stretching and the magmatic history of the margin and the thickness of the sedimentary prism that developed on the margin. Margin-parallel shear may generate thin slivers that may move past their parent continent for thousands of kilometres as parts of other plates (e.g. the Levant fragment west of the Dead Sea Fault). If the continental margin has an irregular shape, this may lead to strike–slip separation of slivers as in the case of the Baja California.

Strike–slip faulting occurring in arcs commonly is placed either along the active volcanic axis (Dewey 1980) or, in less common cases, along the forearc region as in the case of the Atacama fault or both (Katili 1970). Such faulting is most commonly a result of oblique convergence (Fitch 1972), whereby the convergence vector is partitioned into head-on subduction and pure strike–slip behind the frontal (Dewey 1980) or sliver (Jarrard 1986) plate. Coastwise transport (Beck 1988) may move such sliver plates for thousands of kilometres as independent entities. During their transport, they may disintegrate into a number of equant blocks that rotate in harmony with the sense of shear as in Chile (Jeniskey et al. 1987) and in Alaska (Scholl 1989). Alternatively, the strike–slip fault bounding them against the 'arc' may migrate perpendicular to its trace thereby either slicing the sliver plate into ever thinner slivers and stringing it out along the subduction front (E. Irving & P. J. Wynne, this symposium) or adding to it newer bits as it moves along, creating what may be called a 'horizontal stack'. Strike–slip faults that connect those along the magmatic axis (or just behind an accretionary forearc) with the trench may excise or repeat segments of forearcs along cross sections across the convergent zone (Karig 1980; Dewey & Shackleton 1984).

Strike–slip faults in collision zones are much more diverse and complicated individually than those located in other tectonic environments because they affect generally larger areas of the continental lithosphere that has a lower shear strength than oceanic lithosphere and may have been softened further by arc magmatism (Dewey 1980; Smith 1981). Strike–slip faulting generated by continental collosion may be viewed in two broad classes: (1) strike–slip along the suture and (2) strike–slip within the colliding masses that may also cut the suture.

Strike–slip along the suture may be caused by oblique collision (e.g. along the Himalaya) or by change of direction of relative motion upon collision (e.g. along the Zagros). Such strike–slip motion could form complex flakes, negative flowers, or horizontal stacks along the suture zone, considerably complicating its geometry. An extreme case is known from Iran, where the coastwise left-lateral movement of the Podataksasi arc for some 2000 km disrupted the original Palaeo-Tethyan suture zone.

Strike–slip faulting within colliding masses creates large intra-continental transform faults or their shallower intro-lithospheric equivalents, of which the Altin Dagh fault is a particularly fine example. Nearly all of its offset of hundreds of kilometres is taken up by shortening in the

Yorkand arc and in Nan Shan. Very little of it is dispersed into east China to form the Circum-Ordos graben systems. These faults serve to shorten the lithosphere parallel with the convergence direction and lengthen it sideways. If they join with subduction zones elsewhere they then become proper 'escape' avenues, of which the finest example is possibly still the North Anatolian fault in Turkey (Ketin 1948; Sengör 1979).

Strike–slip faults confined to shallower levels than the lithosphere (e.g. crust, upper crust, sedimentary cover, décollement sheets within the sedimentary cover) form abundantly in collision environments, but displacements along them decrease with decreasing penetration. Those that cut through the crust may have displacements on the order of 100 km; those that are confined to levels within the crust can have offsets up to about 100 km; those confined to the sedimentary cover usually have displacements not exceeding about 10 km. Penetration of strike–slip faults is a function of the disruption they cause.

Although there is little doubt that some limited escape of lithospheric and crustal masses occurs in collision belts, its extent seems much less than that maintained by Molnar & Tapponnier (1975) and Tapponnier *et al.* (1980) in southern and eastern Asia. Argand's (1924) old picture of dominant shortening and thickening with not much sideways motion in front of India seems largely vindicated by recent work (Dewey *et al.* 1989).

(iii) *During subduction.* Subduction is an extremely complicated process that creates a wide variety of structures mainly in the upper plate (Dewey 1980; Jarrard 1986). Whether an Andean-type compressive mountain range, or a Mariana-type extensional orogen with numerous marginal basins and remnant arcs, or indeed just a Makran-type 'neutral' arc will form above a given subduction zone seems to depend on whether the trench-line is fixed with respect to a stationary or slowly moving asthenospheric reference frame, and on the date of the roll-back of the hinge-line of the subduction.

In extensional arcs, diverse buoyant slivers may be created by pulling out bits from a pre-existing continental margin in the form of migratory ensialic arcs (e.g. Tyrrhenian–Calabrian arc system, Japan arc system) or by creating ensimatic arcs by successive episodes of marginal basin opening as in the case of the Mariana–Philippine Sea System. In neutral arcs, the most common way of creating buoyant slivers is by slicing the arc by strike–slip faults sub-parallel with its axis. In compressional arcs, the main method of forming fault-bounded geological entities is thrusting on a large scale. Much of this thrusting takes place behind the arc, at elevations lower than about 3 km (Dewey *et al.* 1988). If the subduction zone is steep enough to allow the presence of a mantle wedge beneath the arc (as under Altiplano); a well-developed magmatic arc forms and allows the concentration of shortening in a retroarc fold and thrust belt, whose basal detachment may root into the softened arc core (Dewey 1980). If no mantle wedge is present owing to shallow subduction angle, no magmatic arc develops and shortening becomes distributed in the 'cold edge' of the craton, creating Rocky Mountain-type basement-cored pop-up structures (e.g. the sierras Pampeanas (Jordan & Allmendinger 1986)). The displacements on the thrust faults created along the retroarc region of the compressional arcs are small compared with that along coast-parallel strike–slip faults active in the forearc regions. Both in the Canadian Cordillera and in the Andes, the total retroarc shortening has been estimated to be around 200 km.

In accretionary forearcs, large discontinuities separate distinct packages of mélange that differ from neighbouring packages in terms of matrix composition and age, block composition and age, and structural and metamorphic history. Such discontinuities may result from thrust and normal faulting (Suppe 1972) or strike–slip faulting (Karig 1980) of considerable magnitude (greater than 100 km). Many of the thrust faults in subduction–accretion complexes are mini-sutures that mark lines along which material was accreted to the toe of a growing wedge, although some faults are very difficult to distinguish from those that form by the internal failure of the wedge. Out-of-sequence thrusting and thrust faulting involving the post-accretion inner slope basin strata may be criteria to distinguish accretion faults from wedge failure faults.

Strike–slip faulting, apart from Woodcock's (1986) trench-linked faults, appear in the form of conjugate sets in accretionary complexes. Commonly, one set is dominant and displacement along its members is not significant. The main post-accretion disruption in such complexes appears to be associated either with intra-wedge thrusting or strike–slip faulting related to oblique subduction.

(iv) *During collision.* Collision shortens and thickens the continental crust mainly by thrust faulting and, to a more limited extent, by homogeneous bulk shortening and thickening. As crust thickens, conjugate strike–slip faulting takes an increasingly more significant part in the shortening process by sideways elongation of the shortened region and may accomplish much disruption of pre-existing structures. Escape occurs in front of colliding premontories and towards nearby oceanic pockets, although this process seems less common than hitherto believed.

Major strike–slip faulting in collision orogens may be divided into two classes; those that result from the closure of small oceans and/or where collision was preceded by convergence at high angles to former continental margins, show less disruption by strike–slip faulting than those collisional orogens that grew out of the destruction of large, Pacific-type oceans in which subduction had a variety of orientations with respect to the colliding continental margins before collision. The Alps and the Caledonides provide examples of two such cases respectively.

Orogenic complexity has two main sources: plate kinematics and rheology of the lithosphere. When these two factors are combined, the resultant spectrum of geological processes that generate structures in orogenic belts is theoretically so wide that some of the predicted complexities have not yet been mapped in the field (Dewey 1975). Geological structures in orogenic belts had been catalogued extensively before the rise of plate tectonics. With the appearance of the theory of plate tectonics, the first stage in orogenic research was to accommodate these old observations and the ideas they had given rise to in the new theory and to sieve out or modify those that looked obsolete such as tectogenes and geosynclines. The second present stage was to expand the theory with new observations made in its light. The full range of many of the geological processes that we described above were first recognized during this second stage of orogenic research in the light of plate tectonics.

Terranology manifestly does not advance our understanding of the processes in tectonics unless genetic questions about spatial relationships across discontinuities in orogenic belts are asked during all stages of an investigation. This terranology forbids: 'Where tectono-stratigraphic sequences are not unequivocally correlative, we take a conservative stance and treat them as separate terranes. This taxonomic splitting permits a variety of palinspastic interpretations and reconstruction *once* pertinent data are gathered' (Schermer *et al.* 1984, p. 112, our italics). Here, Schermer *et al.* (1984) allow interpretation only after the pertinent data are gathered but they do not specify how the data are to be gathered; in the light of what question, what theory? What does the 'pertinence' they mention refer to? Pertinent data is a meaningless concept if there is no problem and no theory, in the light of which they can be gathered. Schermer *et al.* (1984, p. 112) further state that 'The scepticism about genetic relationships that is inherent in terrane analysis allows more objective collection and interpretation of data, as facies relationships must be proven before they serve as the basis for a model'. This is unrealistic, because it implies that no model of orogenic evolution could ever be developed as it is not possible to prove all facies relationships in an orogenic belt, if for no other reason than some have already been eroded away.

The implied and advocated scepticism (agnosticism in disguise) allows terranologists to define terranes in the most general and vague terms. Coney *et al.* (1980, p. 330), for example, 'define' the Tracy Arm terrane as follows: 'Structurally complex assemblage of marble, pelitic gneisses and schist of unknown ages.' This description is consistent with almost any environment in an orogenic belt and consequently says very little of boundary relationships, indeed, it says nothing at all. Had Coney *et al.* (1980) given an interpretation of the Tracy Arm assemblage, their readers could have deduced certain predictions from it not contained in the above description and perhaps used them in testing Coney *et al.*'s model and/or used it in a more comprehensive evaluation.

MAP VIEW OF OROGENIC EVOLUTION: IMPORTANCE OF STRIKE–SLIP FAULTING

In this section, we try to answer the question whether terranology has some objective methodology not inherent in other 'methods' of field work, regional geology, data analysis and synthesis by reviewing the methods that led to the widespread recognition of the immense horizontal mobility in orogenic complexes. As the recognition of the importance of strike–slip faulting played a dominant role in this process, we concentrate on the history of the growth of understanding of this particular facet of orogenic evolution.

Although strike–slip faulting at high angles to the trend of orogenic belts had been recognized in the middle of the last century, strike–slip faulting parallel or subparallel with orogenic trend was a 20th-century discovery prompted by the 1906 San Francisco earthquake. It was later incorporated into mobilist theories by Argand (1916, 1924) and Suess (1949) but was not emphasized until an upsurge of interest in strike–slip faulting and its role in orogeny in the 1950s.

Four different approaches led to a revival of interest in strike–slip motion parallel or subparallel with the trend of orogenic belts. The first was the recognition, in the Canadian Cordillera, of anomalous, equatorial ('Tethyan') fossil assemblages of late Palaeozoic age that contrasted with coeval but higher latitude taxa of the cratonic North America. This evidence was used by Wilson (1968) to suggest that parts of the Canadian Cordillera had originated away and farther to the south than where they are now. This implied a path of motion for these pieces, which had a large component parallel with the trend of the Cordillera. Wilson (1968) also drew on Roddick's (1967) evidence to argue for significant strike–slip motion along the Tintina–Rocky Mountain trench system. Later work on 'Tethyan' fossil assemblages (see, for example, Monger & Ross 1971), and especially on palaeomagnetism suggested, in the 80s, that a much larger amount of strike–slip faulting must have occurred during the development of the North American Cordillera.

The second approach has been seismological and involved study of motions along active convergent plate boundaries. McKenzie (1970, 1972) noticed that collision of irregular continental margins leads to disintegration of colliding headlands by sideways expulsion, into Eastern Mediterranean-type oceanic embayments, of lithospheric slivers bounded by conjugate strike–slip faults striking at low angles to the overall trend of the collision zone. Dewey & Burke (1974) pointed out the general applicability of this model to most orogenic belts. The sideways expulsion of material away from colliding promontories was called 'tectonic escape' by Burke & Sengör (1986).

Seismological studies along collision zones also documented widespread shallow (8–25 km) detachments (see, for example, Seeber & Armbruster 1979) that are deformed by strike–slip faults with strikes at low angles to the convergence front. Aric (1981), for example, documented a left-lateral wrench system along the Mur-Mürz-Leitha line in the Austroalpine nappes of eastern Austria forming the southern boundary fault of the thin-skinned Vienna pull-apart basin (Royden 1985). In this case, it is clear that this entire system of wrench fault is confined to the Austroalpine thrust sheets and do not penetrate any deeper.

The third approach has been palaeomagnetic, revealing either by direct demonstration of palaeolatitudinal change motion parallel with roughly meridional orogenic belts such as the Cordillera of North America (see, for example, Beck 1976, 1980; Monger & Irving 1980; Alvarez *et al.* 1980; Cox 1980) or suggesting the incidence of such motion by documenting its effects such as block rotations (see, for example, Beck 1976, 1980; Cox 1980).

The fourth and the most difficult method has been that of traditional field geology combined with some seismic reflection profiling and other geophysical methods such as magnetics. This approach has suggested and documented broad zones of shear along orogenic belts involving substantial rotations (Carey 1958); it also documented orogen subparallel shallow strike–slip faults both in zones of continental collision (see, for example, Trümpy 1977) and in subduction–accretion complexes above subduction zones, deeper, arc-parallel strike–slip faults (Allen 1965; Katili 1970), escape-related strike–slip faults (Ketin 1948; Sengör 1979).

Through a combination of all these approaches a vast amount of strike–slip faulting and associated disruption of orogens creating mosaics of deformation (fig. 10) had been discovered, mapped, and discussed before 1980. There is little doubt that Carey's (1958) prophetic vision was influential in stirring up mobilist interpretations in the Cordillera and reviving it in the Tethyan system. All such interpretations found a ready home in plate tectonics, and, although initially neglected (except by Wilson 1968), they were later incorporated into models of orogenic evolution (see, for example, Dewey *et al.* 1973; Hamilton 1979).

From the foregoing historical review, it seems that there is nothing novel about the methodology of terranology. The answer to the question, therefore, whether terranology has some objective methodology not inherent in other 'methods' of fieldwork, regional geology, data analysis and synthesis, must be no.

Yet, much fuss has been made by terranologists about the methodological superiority of their approach as opposed to former 'classical' plate tectonic models. The following quotations, expressing this view, have been selected at random from the writings of the original promulgators of the concept (all italics ours).

1. '...the procedure of identification of suspect terranes has proven *operationally* very fruitful. It has spawned a completely new type of tectono-stratigraphic map...the compilation of which is providing great insight into Cordilleran tectonic evolution' (Coney 1981, p. 23).

13

2. 'The application of terrane analysis to specific geologic settings, however, produces a *fundamentally different perception of geologic history than that gained through application of plate tectonic "models"*, particularly those dealing with relations along active continental margins...'.

'Tectonic synthesis is the ultimate step in understanding the origin and evolution of continents, and *terrane analysis is the most objective and successful means of achieving this synthesis*' (Jones *et al.* 1983 *b*, p. 103).

3. 'The suspect terrane concept... *provides new insights*.... It is a surgically clean analytical approach and a *superior framework* in which to view the anatomy of any complex orogen' (Williams & Hatcher 1983, p. 33).

In the following and final section, we single out the distinctive aspects of terranology to provide an answer to our final question as to whether terranology has been dangerous or obscuring in any way.

CONCLUSIONS AND DISCUSSIONS: METHODOLOGY OF TERRANOLOGY

For traits of terrane methodology distinct from other, older approaches we must again turn to the statements of the originators of the idea. The central methodological claim of terranologists has been that terrane analysis is an objective, model-independent method of looking at orogenic belts. Schermer *et al.* (1984, p. 112) wrote, for example: 'Therefore, one must be cautious in applying plate tectonics models that assume that tectonic domains that are presently spatially juxtaposed are genetically related. Separation of fact from interpretation is essential here: *one does not define a terrane because it fits or does not fit a model but because of its distinctive stratigraphy and geologic history*.' The terranalogist's own practice belie this statement. In the following, we present a random selection of quotations taken from those sections of various terrane papers devoted to 'terrane characterization' or 'terrane stratigraphy' or 'depositional history' and expressly not from their 'interpretative' sections. The quotations are thus from those parts of these papers that allegedly do not depend on whether they 'fit or (do) not fit a model'.

'Where exposed, older parts of the (Wrangellia) terrane consist dominantly of sedimentary *and arc-related volcanic rocks* that nowhere may be older than Pennsylvanian and, from indirect evidence, may have been, in part, *deposited on oceanic crust*' (Jones *et al.* 1977, p. 2565).

'Upper Triassic and Lower Jurassic marine sandstone and argillite...structurally overlies on the northwest a *dismembered ophiolite (oceanic crust*...' (Jones *et al.* 1977, p. 2571).)

'...have already called attention to the significance of the Chulitna district as possibly representing a *suture zone* along which unlike terranes have been juxtaposed' (Jones *et al.* 1977, p. 2571).

'G, Goodness (composite) – includes three terranes: ...(3) Mesozoic *arc-derived volcanic flows*, tuffs, and greywacke with interbedded chert' (Coney *et al.* 1980, p. 330).

'V, Vizcaino (composite) – includes... Upper Jurassic *arc-derived* volcanic and volcaniclastic rocks...' (Coney *et al.* 1980, p. 331).

'The geologic record of the Chulitna terrane commences with *formation of ocean crust (ophiolite)* in late Devonian time.... Conditions changed markedly after Mississippian time...presumably *this volcanism reflects rifting*...' (Jones *et al.* 1980, pp. A9–A10).

'The characteristic stratigraphic elements of this terrane are (1) an upper Paleozoic *magmatic arc assemblage*...' (Jones *et al.* 1982, p. 3713).

14

'YID Yidun area: (Baiyu–Yidun terrane of Xiong, Zhongza terrane of Zhang *et al.* (1983); may be a *microcontinent caught in subduction complex* of MEK...

'YMU Yushu-Muli: Tr *island arc volcs*, Upper Tr melange; ophiolite and blueschist belt is eastern boundary.

'ZFB Zunggarian foldbelt:...some upper Pz ophiolitic and *island arc rocks...*; *ophiolite possible obducted in late Permiean during collision of island arc...*' (Howell *et al.* 1985, p. 25).

In the above quotations we have italicized those parts of the 'descriptions' that are dependent on plate tectonic interpretations, i.e. 'models'. It is thus clear that, as logically there can be no observation without an *a priori* framework, i.e. a model (see Popper 1968; Feyerabend 1988, p. 63), terrane identification and description are no exceptions.

During this meeting, one of the most distinguished exponents of terrane analysis, David L. Jones, told us in discussion that our criticism on this point was unfair. He pointed out that model-based concepts are of course used to *characterize* a unique sequence of *events*, whose record make up a terrane. He indicated that terranology was really directed against certain specific, individual plate tectonic models that had assumed spatial relations between terranes. This assumption of unwarranted conclusions was what they had thought harmful. If this is so, then terrane analysis may be nothing more than plate tectonic modelling, and neither more nor less model-dependent or objective than any other kind of tectonic modelling. We do not believe, however, the situation to be as advocated by Jones. By its own admission, terranology rose against certain individual plate tectonic models. We can further specify this by stating that it was devised as a refutation of the hypothesis that much of the western North American Cordillera consisted of *in situ* tectonic units, i.e. the present spatial relations of its constituent tectonic units reflect original relations. This may be formulated in a different way as follows: In the western North American Cordillera, there are no tectonic units that were originally so far away from their present locations that they have no genetic links with their present surroundings. This is a universal statement on the scale of the North American Cordillera and, like all natural laws, it prohibits the occurrence of certain events or the presence of certain objects (Popper 1968, pp. 68–69). As such, it has proved to be a very fruitful hypothesis. Terranology rose against this by citing individual examples where this hypothesis failed; it began in Alaska (Berg *et al.* 1972; Jones *et al.* 1972; Jones *et al.* 1972; also cf. Howell 1985*a*) and then spread southwards via Canada (see, for example, Jones *et al.* 1977) to the U.S. and Mexico to engulf the entire Cordillera (see, for example, Coney *et al.* 1980). The individual examples cited all negated the universal statement that 'there are no tectonic units that were originally so far away from their present locations that they have no genetic links with their present surroundings' by showing that there *are* such units. Thus, by nature, these statements all have the form of what Popper (1968, p. 68) calls *strictly existential statements*. The negation of a strictly universal statement (such as a scientific hypothesis) is always equivalent to a strictly existentialist statement and vice versa (Popper 1968, p. 68). Scientific hypotheses are refuted by purely existential statements (also called 'there-is' statements because they refute laws of the form 'there-is-no...'). Terranology thus consists of a large number of purely existential statements and these statements are taken to be 'objective' because they are based on observation (and therefore subjective as every singular observation, unless reproducible, remains subjective: see especially Popper (1968), p. 45 ff). We have, however, seen above the hypothetical character of all observations. But even if we do not consider the hypothetical character of observations, there appears another problem associated with the purely existential statements: They are not refutable because no singular statement can possibly contradict an existential statement. The very idea of 'suspect' terranes, which may be formulated as 'there are (may be) terranes that originated far away from North America, so that they have no genetic ties with their present surroundings' cannot possibly be contradicted by any number of observations because however many 'native' terranes one may find, there may still be those that are exotic. This is so because our observations cannot possibly be exhaustive. Even if we managed to check all of the Cordillera and found no terrane that had come from afar, the possibility of the existence of a now-eroded otherwise-removed one or cryptic terrane can never be excluded, as in many, deeply eroded Precambrian orogenic belts.

Our main methodological criticism of terranology is this property of irrefutability, because of the uncertainty or high probability (in the sense of admitting too much: terranes may be native or exotic, or even cryptic but the theory – if there is such a thing – is non-commutable and therefore cannot be tested) of its formulation. It negates a universal statement, but does not replace it with another one. It stays at the stage of an existentialist statement that is not informative. That the terranes are 'suspect' is not an informative statement, as anyone could suspect anything. In Coney's (1978, p. 38) original definition of suspect terranes, he specifically pointed out that it had been devised so as to avoid questions: 'They are best described as simply 'suspect' terranes. In this way, the question of how far they have travelled is avoided: But science does not avoid questions, it goes after them, because it wants to make the mistakes as far as possible! Scientists do not stay suspicious, but solve problems.

'Terranologists' fear of universal statements (i.e. bold conjectures, hypotheses: 'the terrane concept...is based on an uncertainty principle' (Jones *et al.* 1983*b*, p. 103)) and their tendency towards a much-praised 'in-built scepticism toward plate tectonic modelling between groups of terranes' that characterizes terrane analysis and the demand that 'genetic links must be physically demonstrated between adjacent terranes before palaeogeographic and tectonic models can be constructed using them as building blocks' seems to result from a mistaken view of objectivity. In the natural sciences, objectivity historically has meant, independence of opinions or personal whim and justifiability by anyone who is interested by direct observation. If a statement is to be called 'objective', it has been thought, it has to be understood and *verified* by anyone. The word 'verification' here provides the key to the following discussion.

As Popper (1968, p. 70 footnote 2) points out, if it is characteristic of empirical science to look upon singular statements (such as accounts of observations or experiments) as test-statements, then, with respect to these, universal statements are falsifiable only and existential statements verifiable only. Because objectivity has been thought to imply verifiability by anyone who is interested, good, i.e. objective science, it has been thought, should consist of verifiable, i.e. existential statements. But existential statements, although verifiable on endless individual cases, cannot be tested for universal, i.e. theoretical conclusions, because induction does not work. Thus objectivity of theories of statements pertaining to how the Earth works depends on their testability, for they can neither be verified nor justified by any means (cf. Popper 1968, p. 44 ff). By contrast, observations that are not communicated, must remain subjective and untestable. It is this testability that makes theories independent of anyone's whim, because they can be tested, and if found to be false, discarded. This condition of testability is in fact included in the essence of the word 'objective' that comes from the Latin *objicere* meaning 'to oppose, to stand up to' (Lorenz 1987, p. 17).

16

TERRANOLOGY: VICE OR VIRTUE?

The logical conclusion that existential statements cannot be falsified (i.e. cannot be tested for universal applications) causes terranology to consist of singular existential statements, not testable or much less testable than the plate tectonic models against which it rose. Because terranology is less testable, it becomes consequently less objective. It has been pointed out, however, that the purpose of terrane analysis is not the erection of new universal hypotheses but to show, in individual cases, that the ones so far erected do not work and that a study of orogenic belts must be approached by piecemeal field studies without having to deal with universal hypotheses at every step: 'Conveniently, it (i.e. terrane analysis) alleviates the immediate necessity of incorporating every facet of geologic development in a single all-encompassing model' (Williams & Hatcher 1983, p. 34). But this statement makes 'every facet of geologic development' much less testable, because it has to be looked at from much fewer viewpoints. We believe that this, as a principle, is not a merit but a vice (for it is equivalent to avoiding questions). Of course, there may be instances where some 'facet of geological development' may not be immediately incorporated into a 'single all-encompassing model', but such incorporation is the ultimate aim of our research. We pursue geology to find out how the Earth works and if we lose this ultimate aim from sight in principle, geologizing would deteriorate into an enumeration of singular, unconnected existentialist statements, from which no knowledge, in terms of understanding, can possibly emerge.

It is important to point out that all of what we said above about our methodological reservations is not new. Most philosophers of science after Popper have pointed out all this and they (including Popper) have had some important geologists as predecessors (e.g. Hutton 1794; Gilbert 1886, 1896; Daly 1914). Most terranologists do not follow their own advertised methodology either, most are too good and too experienced as scientists to do so! Like most scientists (except a large number of post-Einstein physicists, however, they too seem to have succumbed to the baconian myth that all 'good' science emerges from observations made by unpolluted minds, and they perhaps fancy that is what one should do to break loose of the fabrications of our minds and get to the 'facts of nature'. We think it is high time that we stop preaching this myth in conference halls and class-rooms.

This myth, and the advocated terrane methodology that clearly grew out of it, has indeed caused some harm. It has made more enumeration of taxonomic entities, i.e. terranes, respectable. Enumeration of geographically named real estate hardly improves our understanding of mountain belts, as was shown by a similar failure of the nappe theory in the Alps. Since Argand, few Alpine geologists had thought in genetic terms. They had, indeed, mapped all their major rock units, and the faults bounding them, but their relations had been inadequately questioned. This danger of using a mindless taxonomy to generate uninformative tectonic maps (of the kind of Coney et al.'s (1980) terrane map) is especially acute in places where there is no tradition of scientific critical thinking. In this proliferation of geographically named, but otherwise totally opaque tectonic units, we see the main danger of the application of terrane analysis.

The word terrane is a lump term for a number of older and more informative non-genetic (block and sliver) and genetic (fragment, nappe, strike–slip duplex, microcontinent, island arc, suture, etc.) terms. Because it is less informative, it is less useful than any of these and also because, historically, the term 'terrane' has had a number of different meanings, it is best avoided. Terrane analysis is neither a new way of looking at orogenic belts, nor a particularly helpful one. Its appeal largely stems from the fact that it takes the responsibility of interpretation of Earth history off the shoulders of the geologist. This, contrary to the common belief of the terranologists, reduces objectivity and makes statements about the tectonics of an area essentially untestable. It has done nothing to advance our understanding of processes in tectonics. We hope that its employment is only temporary and that research on orogenic belts will proceed in the light of daring hypothetical statements that are tested by field studies. This, we believe, was the spirit that led to the rise of plate tectonics.

REFERENCES

Allen, C. R. 1965 Transcurrent faults in continental areas. Continental Drift Symposium. *Phil. Trans. R. Soc. Lond.* A **258**, 82–89.

Alvarez, W., Kent, D. V., Premoli-Silva, I., Schweickert, R. A. & Larson, R. A. 1980 Franciscan Complex limestone deposited at 17° South palaeolatitude. *Bull. geol. Soc. Am.* I **91**, 476–484.

Argand, E. 1916 Sur l'arc des Alpes occidentale. *Eclog. Geol. Helvet.* **14**, 145–191.

Argand, E. 1924 La tectonique de l'Asie. *Congr. Géol. Int.* 13th Sess. (Belgique), vol. 1, pp. 171–172. Liège: Vaillant-Carmanne.

Aric, K. 1981 Deutung krustenseismicher und seismologischer Ergebnisse im Zusammenhang mit der Tektonik des Alpenostrandes. *Sitzb. Österr. Åkad. Wiss., Math.-naturw. Kl. Abt.* I **190**, 235–312.

Atwater, T. 1970 Implications of plate tectonics for the Cenozoic tectonic evolution of western North America. *Bull. geol. Soc. Am.* **1**, 3513–3536.

d'Aubuisson de Voisins, J. F. 1828 *Traité de géognosie* (nouvelle édition, revue et corrigée), vol. 1. Strasbourg: F. G. Levrault. (524 pages.)

Bahat, D. 1983 New aspects of rhomb structures. *J. struct. Geol.* **5**, 591–601.

de la Beche, H. T. 1846 On the formation of the rocks of South Wales and South Western England. *Mem. Geol. Surv. Great Britain Mus. Econ. Geol. Lond.* I, 1–296.

Beck, M. E. Jr 1976 Discordant paleomagnetic pole positions as evidence of regional shear in the western Cordillera of North America. *Am. J. Sci.* **276**, 694–712.

Beck, M. E. Jr 1980 Palaeomagnetic record of plate-margin tectonic processes along the western edge of North America. *J. geophys. Res.* **85**, 7115–7131.

Beck, M. E. Jr 1988 Analysis of Late Jurassic–Recent paleomagnetic data from active plate margins of South America. *J. South Am. Earth Sci.* **1**, 39–52.

Ben-Avraham, Z., Nur, A., Jones, D. L. & Cox, A. 1981 Continental accretion: from oceanic plateaus to allochthonous terranes. *Science, Wash.* **213**, 47–54.

Berg, H. C. *et al.* 1972 Gravina-Nutrotin belt – tectonic significance of an upper Mesozoic sedimentary and volcanic sequence in southern and southeastern Alaska. *US Geol. Surv. Prof. Pap.* 800-D, pp. D1–D24.

Burke, K. & Sengör, A. M. C. 1986 Tectonic escape in the evolution of the continental crust. In *Reflection seismology: the continental crust* (ed. M. Barazangi), pp. 41–53, Geodynamics Ser., vol. 14. Washington, D.C.: A. G. U.

Carey, S. W. 1958 The tectonic approach to continental drift. In *Continental drift, a symposium* (ed. S. W. Carey), pp. 177–356. University of Tasmania, Hobart.

Chang, C. F. *et al.* 1986 Preliminary conclusions of the Royal Society and Academia Sinica 1985 geotraverse of Tibet. *Nature, Lond.* **323**, 501–507.

Coney, P. J. 1978 Mesozoic–Cenozoic Cordilleran plate tectonics. *Geol. Soc. Am. Mem.* **152**, 33–50.

Coney, P. J. 1981 Accretionary tectonics in western North America. *Arizona Geol. Soc. Dig.* **14**, 23–37.

Coney, P. J. 1989 Structural aspects of suspect terranes and accretionary tectonics in western North America. *J. struct. Geol.* **11**, 107–125.

Coney, P. J., Jones, D. L. & Monger, J. W. 1980 Cordilleran suspect terranes. *Nature, Lond.* **288**, 329–332.

Cox, A. 1980 Rotation of microplates in western North America. *Geol. Ass. Can. Spec. Pap.* **20**, 305–321.

Daly, R. A. 1914 *Igneous rocks and their origin.* New York: McGraw-Hill. (563 pages.)

Dana, J. D. 1880 *Manual of geology.* New York: Ivison, Blakeman, Taylor and Co. (911 pages.)

De Long, S. E., Dewey, J. F. & Fox, P. J. 1977 Displacement history of oceanic fracture zones. *Geology* **5**, 199–202.

Dewey, J. F. 1975 Finite plate evolution, some implications for the evolution of rock masses on plate margins. *Am. J. Sci.* A **275**, 268–284.

Dewey, J. F. 1976 Ancient plate margins: some observations. *Tectonophysics* **33**, 379–385.

Dewey, J. F. 1980 Episodicity, sequence and style at convergent plate boundaries. *Geol. Ass. Can. Spec. Pap.* **20**, 553–573.

Dewey, J. F. & Burke, K. 1974 Hot-spots and continental break-up: implications for collision orogeny. *Geology* **2**, 57–60.

Dewey, J. F. & Shackleton, R. M. 1984 A model for the evolution of the Grampian tract in the early Caledonides and Appalachians. *Nature, Lond.* **312**, 115–121.

Dewey, J. F., Cande, S. C. & Pitman, W. C. 1989 Tectonic evolution of the India–Eurasia covergent zone. *Eclog. Geol. Helv.* **82**, 717–734.

Dewey, J. F., Pitman, W. C., Ryan, W. B. & Bonnin, J. 1973 Plate tectonics and the evolution of the Alpine System. *Bull. Geol. Soc. Am.* **84**, 3187–3190.

Dewey, J. F., Shackleton, R. M., Chang, Chengfa & Sun, Yi Yin 1988 The tectonic evolution of the Tibetan Plateau. *Phil. Trans. R. Soc. Lond.* **327**, 379–413.

Dufrénoy, A. & de Beaumont, E. 1841 *Explication de la Carte Géologique de la France.* Paris: Roale. (823 pages.)

England, P. C. & Houseman, G. A. 1988 The mechanics of the Tibetan Plateau. *Phil. Trans. R. Soc. Lond.* **326**, 301–319.

Feyerabend, P. 1988 *Against method* (rev. edn). London: Verso. (296 pages.)

Fitch, T. J. 1972 Plate convergence, transcurrent faults, and internal deformation adjacent to southeast Asia and the western Pacific. *J. geophys. Res.* **77**, 4432–4460.

Früh, J.-J. 1888 Beiträge zur Kenntniss der Nagelfluh der Schweiz. *Gekrönte Preisschrift: Denkschr. Schw. Naturf. Gesell.*, vol. 30.

Gary, M., Bates, R. L. & Jackson, J. A. (eds) 1972 *Glossary of geology.* Washington, D.C.: America Geological Institute. (805 + 52 pages.)

Gilbert, G. K. 1886 The inculcation of scientific method by example. *Am. J. Sci.* 3rd ser. **31**, 284–299.

Gilbert, G. K. 1896 The origin of hypotheses: Presidential Address. Washington, D.C.: The Geological Society of Washington. (24 pages.)

Grambling, J. A., Williams, M. L. & Mawer, C. K. 1988 Proterozoic tectonic assembly of New Mexico. *Geology* **16**, 724–727.

Guo, L. 1984 On terrane – a latest concern in the study of plate tectonis. *Bull. Chinese Acad. Geol. Sci.* **10**, 27–34.

Hamilton, W. 1979 Tectonics of the Indonesian region. *Prof. Pap. U.S. geol. Surv.* **1078**, 1–345.

Hashimoto, M. & Uyeda, S. (eds) 1983 Accretion Tectonics in The Circum-Pacific Regions. Tokyo: Terra Scientific. (358 pages.)

Haug, E. 1907 *Traité de géologie*, vol. 1. Paris: Librairie Armand Colin. (538 pages.)

Haug, E. 1925 Contribution à une synthèse stratigraphique des Alpes Occidentales. *Bull. Soc. Géol. France*, sér. 4, **25**, 97–244.

Helwig, J. E. 1974 Eugeosynclinal basement and a collage concept of orogenic belts. *Soc. Econ. Pal. Min. Spec. Publ.* **19**, 359–376.

Howell, D. 1985*a* Terranes. *Scient. Am.* **253**, 90–103.

Howell, D. G. (ed.) 1985*b* Tectonostratigraphic terranes of the Circum-Pacific region. Circum-Pacific Council for Energy and Mineral Resources, Earth Science Series, no. 1. (585 pages.)

Howell, D. G. 1989 *Tectonics of suspect terranes.* London: Chapman and Hall.

Howell, D. G. *et al.* 1985 Tectonostratigraphic terranes of the Circum-Pacific region. In *Tectonostratigraphic terranes of the Circum-Pacific region* (ed. D. G. Howell), pp. 3–30. Houston, Texas: Circum-Pacific Council for energy and Mineral Resources.

Hutton, D. H. W. & Dewey, J. F. 1986 Palaeozoic terrane accretion in the Western Irish Caledonides. *Tectonics* **5**, 1115–1124.

Jarrard, R. D. 1986 Relations among subduction parameters. *Rev. Geophys.* **24**, 217–284.

Jia, C. G. 1988 *Plate tectonics of Eastern Qinling Mountains of China.* (In Chinese.) (130 pages.)

Jones, D. L., Silberling, N. J. & Nelson, W. H. 1972 Southeastern Alaska – a displaced continental fragment? *US Geol. Surv. Prof. Pap.* **800-B**, B211–B217.

Jones, D. L., Silberling, N. J. & Hillhouse, J. W. 1977 Wrangellia – A displaced terrane in northwestern North America. *Can. J. Earth Sci.* **14**, 2565–2577.

Jones, D. L., Silberling, N. J., Csejtey, B., Nelson, W. H. & Blome, C. D. 1980 Age and structural significance of ophiolite and adjoining rocks in the Upper Chulitna District, South-central Alaska. *US Geol. Surv. Prof. Pap.* **112-A**, 1–21.

Jones, D. L., Silberling, N. J., Berg, H. G. & Plafker, G. 1982 Character, distribution, and tectonic significance of accretionary terranes in the Central Alaska Range. *J. geophys. Res.* **87**, 3709–3717.

Jones, D. L., Howell, P. G., Coney, P. J. & Monger, J. W. 1983*a* Recognition, character and analysis of tectonostratigraphic terranes in western North America. In *Accretion tectonics in the Circum-Pacific regions* (ed. M. Hashimoto & S. Uyeda), pp. 21–35. Tokyo: Terra Scientific.

Jones, D. L., Howell, P. G., Coney, P. J. & Monger, J. W. 1983*b* Recognition, character and analysis of tectonostratigraphic terranes in western North America. *J. Geol. Educ.* **31**, 295–303.

Jordan, T. E. & Allmendinger, R. W. 1986 The Sierras Pampeanas of Argentina: a modern analogue of Rocky Mountain foreland deformation. *Am. J. Sci.* **286**, 737–764.

Katili, J. A. 1970 Large transcurrent faults in Southeast Asia with special reference to Indonesia. *Geol. Rundschau* **59**, 581–600.

Ketin, I. 1948 Über die tektonisch-mechanischen Folgerungen aus den grossen anatolischen Erdbeben des letzten Dezenniums. *Geol. Rundschau* **36**, 77–83.

Kerr, R. A. 1980 The bits and pieces of plate tectonics. *Science, Wash.* **207**, 1059–1061.

Lamb, S. 1987 A model for tectonic rotation about a vertical axis. *Earth planet. Sci. Lett.* **84**, 75–86.

Leitch, E. C. & Scheibner, E. (eds) 1987 Terrane accretion and orogenic belts. *Geodynamics Ser.*, vol. 19. Washington, D.C.: A. G. U. (343 pages.)

Lorenz, K. 1987 *Die Rückseite des Spiegels. Versuch einer Naturgeschichte menschlichen Erkennens.* München: Deutscher Taschenbuch Verlag. (318 pages.)

Lugeon, M. 1902 Les grandes nappes de recourement des Alpes de Chablai et de la Suisse. *Bull. Soc. géol. Fr.* **1**, 723–825.

McKenzie, D. P. 1970 Plate tectonics of the Mediterranean region. *Nature, Lond.* **226**, 239–243.

McKenzie, D. P. 1972 Active tectonics of the Mediterranean region. *Geophys. Jl R. astr. Soc.* **30**, 109–185.

McKenzie, D. P. & Jackson, J. 1983 The relationship between strain rates, crustal thickening, palaeomagnetism, finite strain and fault movements within a deforming zone. *Earth planet. Sci. Lett* **65**, 182–202.

McKenzie, D. P. & Morgan, W. J. 1969 Evolution of triple junctions. *Nature, Lond.* **224**, 125–133.

McKerrow, W. S. & Gibbons, W. 1988 Displaced terranes in Britain and Ireland – a joint meeting of the British Terranes Research Group and the Geological Society of London Stratigraphic Committee. *Geol. Soc. Newsletter* **17** (5), p. 12.

Mason, R. 1988 Did the Iapetus Ocean really exist? *Ìeology* **165**, 823–826.

Molnar, P. & Tapponnier, P. 1975 Cenozoic tectonics of Asia: effects of a continental collision. *Science, Wash.* **189**, 419–426.

Monger, J. W. H. 1975 Correlation of eugeosynclinal tectono-stratigraphic belts in North American Cordillera. *Geosc. Can.* **2**, 4–9.

Monger, J. W. H. & Francheteau, J. (eds) 1987 *Circum-Pacific orogenic belts and evolution of the Pacific Ocean basin.* *Geodynamics Ser.*, vol. 18. Washington, D.C.: A. G. U. (165 pages.)

Monger, J. W. H. & Irving, E. 1980 Northward displacement of north-central British Columbia. *Nature, Lond.* **285**, 289–294.

Monger, J. W. H. & Ross, C. A. 1971 Distribution of Fusalinaceans in the western Canadian Cordillera. *Can. J. Earth Sci.* **8**, 259–278.

Moores, E. M. 1970 Ultramafics and orogeny, with models of the U.S. Cordillera and the Tethys. *Nature, Lond.* **228**, 837–842.

Nur, A. 1983 Accreted terranes. *Rev. Geophys. Space Phys.* **21**, 1779–1785.

Nur, A. & Ben-Avraham, Z. 1982a Oceanic plateaus, the fragmentation of continents and mountain building. *J. geophys. Res.* **87**, 3644–3661.

Nur, A. & Ben-Avraham, Z. 1982b Displaced terranes and mountain building. In *Mountain building processes* (ed. K. J. Hsü), pp. 73–84. London: Academic Press.

Nur, A. & Ben-Avraham, Z. 1988 Accreted terranes and the enigma of the Andes. In *Provisional Proc. 4th Int. Tectonostratigraphic Terrane Conf., Univ. Nanjint, Nanjing, China* (ed. D. G. Howell & T. J. Wiley), p. 76.

Pindell, J. & Dewey, J. F. 1982 Permo-Triassic reconstruction of Western Pangea and the evolution of the Gulf of Mexico/Caribbean region. *Tectonics* **1**, 179–211.

Popper, K. R. 1968 *The logic of scientific discovery.* New York: Harper and Row. (480 pages.)

Roddick, J. A. 1967 Tintina trench. *J. Geol.* **75**, 23–33.

Roeder, D. H. 1973 Subduction and orogeny. *J. geophys. Res.* **78**, 5005–5024.

Royden, L. H. 1985 The Vienna basin: a thin-skinned pull-apart basin. In *Soc. Econ. paleont. Min. Spec. Publ.* **37**, 317–338.

Saleeby, J. B. 1983 Accretionary tectonics of the North American Cordillera. *A. Rev. Earth planet. Sci.* **15**, 45–73.

Schardt, H. 1893 Sur l'origine des Alpes du Chablais et du Stockhorn, en Savoie et an Suisse. *C.r. hebd. Séanc. Acad. Sci., Paris* **117**, 707–709 (errata p. 874).

Schermer, E. R., Howell, D. C. & Jones, D. L. 1984 The origin of allochthonous terranes: perspective on the growth and shaping of continents. *A. Rev. Earth planet. Sci.* **12**, 107–131.

Seeber, L. & Armbruster, J. 1979 Seismicity of the Hazara Arc in northern Pakistan: Decollement vs. basement faulting. In *Geodynamics of Pakistan* (ed. A. Farah & K. A. DeJong), pp. 131–142. Quetta: Geological Survey of Pakistan.

Sengör, A. M. C. 1979 The North Anatolian transform fault: its age, offset and tectonic significance. *J. geol. Soc. Lond.* **136**, 269–282.

Sengör, A. M. C. 1990a Plate tectonics and orogenic research after 25 years: a Tethyan perspective. *Tectonophysics.*

Sengör, A. M. C. 1990b Lithotectonic terranes and the plate tectonic theory of orogeny: a critique of the principles of terrane analysis. In *Proc. 4th Int. Circum-Pacific terrane conf., Nanjing* (ed. T. Wiley & D. G. Howell).

Sengör, A. M. C. 1990c A new model for the Late Palaeozoic–Mesozoic tectonic evolution of Iran and surrounding regions and implications for Oman. *Geol. Soc. Lond. Spec. Publ.*

Sengör, A. M. C., Altiner, D., Cin, A., Ustaömer, T. & Hsü, K. J. 1990 Origin and assembly of the Tethyside orogenic collage at the expense of Gondwana-Land. In *Gondwana and Tethys. Proc. First Lyell Meeting* (ed. M. G. Audley-Charles & A. Hallam).

Silberling, N. J. & Jones, D. L. (eds) 1984 Lithotectonic terrane maps of the North American Cordillera. *US Geol. Surv. Open-File Rep.* 84–523. (99 pages.)

Smith, A. G. 1981 Subduction and coeval thrust belts, with particular reference to North America. *Geol. Soc. Lond. Spec. Publ.* **9**, 111–124.

Stephens, M. B. & Gee, D. G. 1985 A tectonic model for the evolution of the eugeosynclinal terranes in the central Scandinavian Caledonides. In *The Caledonide Orogen – Scandinavia and related areas* (ed. D. G. Gee & B. A. Sturt), pp. 954–978. Chichester: Wiley.

Studer, B. 1834 *Geologie der Westlichen Schweizer Alpen.* Heidelberg.

Suess, E. 1875 *Die Entstehung der Alpen.* W. Braunmüller Wien. (168 pages.)

Suess, E. 1891 Die Brüche des östlichen Afrika. *Denkschr. k. Akad. Wiss., math.-nat. Cl.* **63**, 555–584.

Suess, E. 1949 Bausteine zu einem System der Tektogenese. III. Der Bau der Kaledoniden und die Schollendrift im Nordatlantik. B. Die Kaledoniden in Skandinavien. C. Die Kaledoniden in Grönland. *Mitt. Geol. Gessell. Wien* **36–38**, 29–230.

Suppe, J. 1972 Interrelationships of high-pressure metamorphism, deformation and sedimentation in Franciscan tectonics. *U.S.A. Int. Geol. Congr. 24th Sess. Sec. 3*, pp. 552–559.

Tapponnier, P., Peltzer, G. & Armijo, R. 1986 On the mechanics of the collision between India and Asia. *Geol. Soc. Lond. Spec. Publ.* **19**, 115–157.

Tollmann, A. 1973 Grundprinzipien der alpinen Deckentektonik: Wien, Franz Deuticke, 404 p.

Tollmann, A. 1987 Neue Wege in der Ostalpengeologie und die Beziehungen zum Ostmediterran. *Mit. Öster. Geol. Gessell.* **80**, 47–113.

Trümpy, R. 1955 Wechselbeziehungen Zwischen Paläogeographie und Deckenbau: Vierteljschr. *Naturforsch. Gessell. Zürich* **100**, 217–231.

Trümpy, R. 1960 Palaeotectonic evolution of the Central and Western Alps. *Bull. geol. Soc. Am.* **71**, 843–908.

Trümpy, R. 1977 The Engadine Line: a sinistral wrench fault in the Central Alps. *Mem. Geol. Soc. China* **2**, 1–12.

Weber, K. 1986 The mid-European Varisides in terms of allochthonous terrains. In *Proc. Third Workshop on the European Geotraverse (EGT) Project* (ed. R. Freeman, S. Mueller & P. Giese), pp. 73–81. Strasbourg: ESF.

White, R. S., Spence, G. D., Fowler, S. R., McKenzie, D. P., Westbrook, G. K. & Bowen, A. N. 1987 Magmatism at rifted continental margins. *Nature, Lond.* **330**, 439–444.

Williams, H. & Hatcher, R. D. Jr 1982 Suspect terranes and accretionary history of the Appalachian orogen. *Geology* **10**, 530–536.

Williams, H. & Hatcher, R. D. Jr 1983 Appalachian suspect terranes. *Geol. Soc. Am. Mem.* **158**, 33–53.

Wilson, J. T. 1966 Some rules for continental drift. *R. Soc. Can. Spec. Publ.* **9**, 3–17.

Wilson, J. T. 1967 Some implications of new ideas on ocean-floor spreading upon the geology of the Appalachians. *R. Soc. Can. Spec. Publ.* **10**, 94–99.

Wilson, J. T. 1968 Static or mobile earth: the current scientific revolution. *Am. Phil. Soc. Proc.* **112**, 309–320.

Discussion

A. H. F. ROBERTSON (*Grant Institute of Geology, University of Edinburgh, U.K.*). Surely the terrane concept has in fact proved to be very useful, particularly in that it has led to more emphasis on the nature of contacts and the possibility of major strike–slip displacements in many orogenic belts?

J. F. DEWEY, F.R.S. Yes, these are indeed the only two ways in which the terrane 'concept' has been useful.

Synopsis of late Palaeozoic and Mesozoic terrane accretion within the Cordillera of western North America

By D. L. Jones

Department of Geology and Geophysics, University of California, Berkeley, California 94720, U.S.A.

Establishing the paleogeographic origin of most of the terranes within the Cordillera remains an ellusive goal; despite more than 10 years of multidisciplinary research, the home port of any major terrane has not been identified unequivocally. Even most continental fragments that show affinities to North America cannot be repositioned confidently along the Cordilleran margin, and some continental fragments (e.g. Chulita) probably are not North American in origin. Cordilleran oceanic terranes, including island arcs, seamounts, off-ridge islands, and scraps of ocean basins, are especially difficult to reposition because Panthalassa has been destroyed. Faunal studies with emphasis on palaeobiogeographic affinities are the most useful, particularly when coupled with analyses of faunal diversity and endemism. Such studies suggest that some terranes previously thought to have formed near the Cordillerran margin were situated thousands of kilometres to the west, and were separated from the continent by broad ocean basins, rather than by a narrow marginal sea.

Introduction

Beginning in late Permian time and continuing into the mid-Cenozoic, new continental crust was formed along the ancient rifted western margin of North America by accretion of oceanic island arcs, seamounts, pelagic and volcaniclastic sedimentary rocks, together with detached continental fragments and thick prisms of continentally derived sediments. The place of origin for most of the slivers, blocks, and fragments of accreted material remains undetermined, although several palaeomagnetic and palaeobiogeographic analyses document tectonic transport of thousands of kilometres. To emphasize the fact that great uncertainties still exist regarding palaeogeographical relations through time, the term 'terrane' (or 'suspect terrane') has been applied to these fault-bounded, displaced blocks. This uncertainty in place of origin is also reflected in uncertainty in genetic linkage between terranes, and between terranes and the continental margin. A basic tenet of terrane analysis is that mere propinquity is insufficient grounds for establishing genetic linkage in the absence of more compelling arguments.

The use of the term 'terrane' within the context of Cordilleran tectonic development has been clear and relatively straightforward. The term refers specifically to fault-bounded geologic entities of regional extent that differ in significant ways from their neighbours. Such differences between terranes imply relative tectonic dislocations sufficiently large that facies connections are broken and original palaeogeographic relations obscured. Minimum nominal displacements of a few hundred kilometres appear to be required, but in fact, most displacements probably were much larger.

Several misconceptions and complaints have recently arisen regarding terrane concepts and terminology. The most serious objections have been voiced by Dr Sengör (this symposium), and result from lack of understanding of the original terrane literature. Sengör's mistaken view is that terrane analysis ignores the genetic significance of lithotectonic assemblages and instead, is concerned only with simple descriptions of the rocks present without regard to their mode of origin. If such allegations were indeed true, terrane analysis long ago would have been widely rejected as a meaningless exercise. But of course, his complaints are completely unfounded because establishing the genetic significance of lithic assemblages within terranes, and discovering the genetic linkage between terranes, are now and always have been the ultimate goals of terrane analysis.

Another more trivial complaint (see, for example, W. B. Hamilton, this symposium) is that names of terranes are not genetic; that is, no clue as to the genesis of the rocks present is afforded by the name itself. This forces the reader to learn something of the internal composition of the terrane before the name becomes meaningful. In one sense, this is a legitimate complaint, but it is also founded on a lack of understanding of the nature of most terranes within the Cordillera. The basic reason why a non-genetic nomenclature was adopted from the beginning is because most terranes are polygenetic and cannot be described in simple terms. A good example of a polygenetic terrane is Chulitna, a small terrane located in southern Alaska. This terrane is of mixed oceanic and continental character that began with formation of oceanic crust in late Devonian time. Island arc volcanics and volcaniclastics covered the basal ophiolite by late Carboniferous time, to be followed in the Permian to early Triassic by shelfal carbonates and clastics. A major continental collision followed by rifting occurred in late Triassic time as evidenced by the mixture of ophiolitic and polycrystalline quartz and mica schist detritus in Triassic redbeds intercalated with basalt flows. By late Mesozoic time, this entire assemblage had subsided again and was covered by Upper Jurassic and Cretaceous radiolarian chert and flysch deposits. The specific sequence of tectonic events recorded in the sedimentary and structural record of Chulitna is unique to that terrane, but the complexity of events and the rapid change from one tectonic setting to another through time is characteristic of most of the Cordilleran terranes. This complexity precludes use of simple genetic terms to describe the entire sequence.

LATE PALAEOZOIC AND MESOZOIC TECTONIC DEVELOPMENT OF THE CORDILLERA

The main accretionary history of the Cordillera can be subdivided in six diachronous phases that overlap in space and time, but which generally are older to the south than to the north. These phases are summarized below and described in more detail in the following section.

1. Obduction of oceanic terranes, including Golconda, Slide Mtn, Tozitna, and Angayucham, onto the continental margin; time of emplacement ranges from probably later Permian in western Nevada (Sonoman orogeny) to early Cretaceous in the Brooks Range of northern Alaska.

2. Accretion of Asiatic (i.e. Tethyan) oceanic carbonate platform (Cache Creek terrane) in British Columbia in Middle Jurassic time.

3. Accretion of major inner Palaeozoic island arc terranes, including Stikinia, Eastern Klamath, Northern Sierra, and Black Rock, in Middle Jurassic time with resulting deformation of the Upper Triassic 'mudpile' and redeformation of Golconda terrane in northwestern Nevada.

4. Accretion of major outer Palaeozoic and Mesozoic arc terranes in late Jurassic to mid-

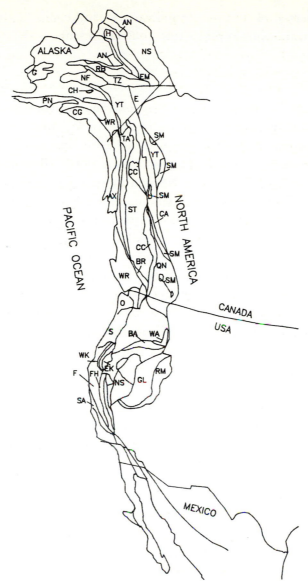

FIGURE 1. Sketch map showing selected accreted terranes in the cordillera of western North America. Symbols are as follows. Alaska: AN, Angayucham; CG, Chugach and McHugh; CH, Chulitna; E, Eagle; EM, Endicott Mountains; G, Goodnews; H, Hammond; NF, Nixon Fork; NS, North Slope; PN, Peninsular; RB, Ruby; TZ, Tozitna; WR, Wrangellia; YT, Yukon Tanana. Canada and SE Alaska: AX, Alexander; BR, Bridge River; CA, Cassiar; CC, Cache Creek; QN, Quenellia; SM, Slide Mountain; ST, Stikinia; TA, Tracy Arm; YT, Yukon Tanana. U.S.A.: BA, Baker and Izee; EK, Eastern Klamath; F, Franciscan and Great Valley sequence; FH, Foothills; GL, Golconda; NS, Northern Sierra; O, Olympic; RM, Roberts Mountain; S, Siletz; SA, Salinia; WA, Wallowa; WK, Western Klamath.

Cretaceous, including Wrangellia in British Columbia and Alaska, Wallowa, Huntinton, and Baker terranes in eastern Oregon, and Western Klamath and Foothills terranes in western Oregon and central California.

5. Development of integrated obliquely east-dipping subduction system along the entire Pacific Coast in Cretaceous time with accretion of graywacke and melange prisms from Baja California north to southern Alaska, including the Franciscan and Pacific Rim Complexes, and the Chugach and McHugh terranes.

25

6. Northward dispersal of terrane fragments by strike–slip faulting within accreted continental crust; continuous from mid-Cretaceous to present.

PHASE 1. OBDUCTION OF OCEANIC TERRANES

The first accretionary events along the continental margin consisted of obduction of vast sheets of radiolarian chert, argillite, mafic volcanic rocks, and minor volcaniclastic and siliciclastic rocks. Age-span of obducted rocks is late Devonian to late Permian in the south (Nevada) and late Devonian to early Jurassic in the north (Brooks Range), and time of emplacement varies from late Permian in Nevada to early Cretaceous in the Brooks Range. Rock assemblages are generally similar throughout, although local variations in the percentage of rock types present are notable. Abundance of mafic volcanic rocks is particulary variable, with some terranes being nearly volcanic-free, whereas others, such as Angayucham, being composed dominantly of basalt and related mafic intrusive rocks. Although many of these mafic volcanic assemblages have been referred to as 'ophiolite', geologic associations and chemical data indicate that the bulk of them represent seamounts or scraps of oceanic plateaus, rather than primary oceanic crust. Seamount volcanism in Alaska occurred in late Devonian, Carboniferous, Permian, and Triassic times, indicating that the oceanic domain in which the seamounts formed remained tectonically active (rifting?) for nearly 200 Ma.

These oceanic assemblages have been variously interpreted as having formed in a back-arc basin and to have been obducted by 'back-arc thrusting'; or to have formed in an open ocean, and to have been offscraped to form an accretionary prism above a west-dipping subduction zone into which the continental margin migrated. Despite the long record of pelagic sedimentation and associated volcanic activity found throughout the oceanic terranes, no compelling evidence has yet been presented that an active arc persisted on the west side of the oceanic basin, or that internal deformation or accretion took place during a protracted period of time. Thus neither model is very attractive in that both fail critical geologic tests. A narrow back-arc basin seems precluded by the long duration of oceanic sedimentation and by the large number of separate tectonostratigraphic environments represented within the terranes. These differing environments include continental rise and slope deposits, seamount provinces, pelagic basins free of volcanic activity, pelagic basins with arc-derived volcaniclastics, and limestone turbidite fans that lack obvious sources. Emplacement of the oceanic thrust sheets within a broadly transpressive transform system seems more probable than either 'back-arc thrusting' or offscraping above a normal subduction zone. Evidence for both left-lateral and right-lateral transpression are locally compelling, but the palaeogeography and mode of emplacement of the oceanic terranes remain as major unresolved problems.

PHASE 2. ACCRETION OF CARBONATE PLATFORM

The well-known Cache Creek terrane of northern British Columbia consists of a thick coherent sequence of shallow water carbonate rocks deposited on a basaltic substratum (seamount or oceanic plateau). Dated rocks range in age from early Carboniferous to Permian, with distinctive Asiatic (i.e. Tethyan) fusulinids characteristically occurring in the Permian part of the section. Farther south in British Columbia, the sequence is more disrupted as Permian and early Triassic limestone occurs as large blocks floating in a Triassic or early

Jurassic chert-argillite melange. Elsewhere, in Washington, Oregon, California, and Alaska, small blocks of limestone with Tethyan fusulinids have been identified as 'Cache Creek' terrane, although without exception, these blocks occur in Jurassic or younger melanges whose relations to the type Cache Creek terrane are not clear.

Based on the nearly total dissimilarity in Permian faunas between the Cache Creek terrane and the North American continental margin, and the strong similarity of Cache Creek fusulinids and corals to those occurring in Japan, China, and other parts of Asia, it is reasonable to propose that the Cache Creek terrane originated in an oceanic domain in the western part of Panthalassa beyond the effective range of North America faunal migrants. A minimum tectonic displacement of Cache Creek terrane comparable with the present equatorial width of the Pacific Ocean does not seem excessive.

PHASE 3. ACCRETION OF INNER ARCS

The first major episode of accretion of Palaeozoic island arcs occurred in middle Jurassic time, and brought Stikinia, the Eastern Klamath (including the small scrap at Bilk Creek in northwestern Nevada), and the Northern Sierra terranes into contact with the North American margin. Convergence must have occurred on an obliquely west-dipping subduction system, as arc volcanism did not effect most of the continental margin at this time. The main results of the collision were metamorphism of the Ominica crystalline belt in Canada and widespread thrusting, metamorphism, and crustal thickening in Nevada and eastern California. Much of the previously deformed Golconda terrane of western Nevada was redeformed at this time, together with Upper Triassic and Jurassic 'overlap' deposits that are younger than the Sonoman orogeny. The relations of the Eastern Klamath terrane and the Northern Sierra terrane remain controversial. Although many authors treat the two as parts of a single arc system, geochemical data and differences in stratigraphy support their separation until at least late Triassic time.

PHASE 4. ACCRETION OF OUTER ARCS

Following accretion of the major Palaeozoic arc terranes to the continental margin in middle Jurassic time, another assemblage of arcs of both Palaeozoic and Mesozoic age was accreted in later Jurassic to mid-Cretaceous time. Attendant to their accretion, a complex assemblage of deep water flysch deposits were crushed between the converging terranes and the continental margin. In Alaska and British Columbia, the primary terranes accreted during the Cretaceous comprise the amalgamated super terrane composed of Wrangellia, Alexander terrane, and the Peninsular terrane. Driven northward ahead of this amalgamated massive terrane was a series of smaller scraps and fragments of mixed oceanic, island arc, and continental character, including Chulitna terrane, that finally lodged within upper Mesozoic flysch belts of southern and northwestern Alaska.

The accretionary record of Wrangellia is clearly expressed in its late Mesozoic sedimentary and structural history. The first and only period of major folding, thrust faulting, and deep erosion within Wrangellia occurred between late Jurassic and mid-Cretaceous (Albian) time. Several complicated pulses of deformation can be recognized during this interval, but the important point is that a major reorganization of the entire depositional system occurred at this time, which is believed to mark the first collision of Wrangellia with the continental margin.

Following this collision subsidence commenced in the Albian with development of local deep basins; this subsidence continued without obvious interruption until near the end of the Cretaceous when strong uplift occurred and marine deposition was halted.

In contrast to the deformational history of Wrangellia which records an important pre-Albian collision, the surrounding flysch basins were deformed during the late Cretaceous. This difference in age suggests that the original collision of Wrangellia and its final emplacement into Alaska were separate events, the former having occurred somewhere to the south.

Accretion of outer arcs in California and eastern Oregon occurred mainly in the late Jurassic, an age broadly equated with the Nevadan orogeny. In both the Foothills and the Western Klamath terranes Upper Jurassic arc volcanics and associated epiclastic sedimentary rocks structurally overlying a complex oceanic basement were deformed and accreted to the continental margin in Kimmeridgian time. Volcanic rocks are predominantly augite porphyries, but dacitic to rhyolitic rocks are locally abundant. Relations of the volcanic rocks to underlying ultramafic to mafic to intermediate intrusive and extrusive rocks and associated sedimentary rocks of Upper Palaeozoic to middle Jurassic age are obscure, and several differing interpretations have been put forward. In the northern part of the Foothills terrane, Jurassic arc volcanics and associated plutonic rocks are interpreted to have been thrust eastward over the basement assemblage above a west-dipping subduction zone. In the central Foothills terrane, some workers interpret the arc volcanics to have been deposited above the basement rocks as a consequence of an east-dipping subduction zone that extended beneath the North American continental margin. Whatever the original disputed relations may have been, the present structural relations between the Upper Jurassic arc volcanics and the basement rocks appear to consist primarily of low-angle extensional faults of regional extent that have been folded by later compressive events associated with the terminal phases of the Nevadan orogeny. This postulated period of major extension may be partly responsible for generating large bodies of serpentinite melange that characterizes large tracts of the basement.

In eastern Oregon, the Wallowa terrane was emplaced in Cretaceous time against Mesozoic plutonic rocks intruded into Precambian basement. Earlier accreted rocks, as well as Paleozoic miogeoclinal strata, are missing in the suture zone, implying that major truncation of the continental margin occurred before the final accretion of Wallowa terrane.

PHASE 5. ACCRETION OF UPPER MESOZOIC AND CENOZOIC SUDDUCTION COMPLEXES

By the end of Jurassic time in the south, and by the end of Cretaceous time in the north, the oceanic domains bordering North America had been swept clean of Palaeozoic and Mesozoic island arcs and other terranes with thick crustal sections. Unimpeded subduction of oceanic crust beneath the continental margin was now possible for the first time in a belt extending from Baja California north to southern Alaska. Although most earlier interpretations of this integrated subduction system emphasized normal, or head-on, convergence between oceanic plates and the continent, abundant palaeomagnetic and faunal data now show that convergence was highly oblique with a northward component that was at least comparable in magnitude with the east–west component of convergence. The result of this obliquity was the internal disruption of the fore-arc and back-arc regions by right lateral strike–slip faults that have rearranged and modified the original palaeogeographic relations within the entire region of the continental margin. Well-known accretionary prisms that formed during this phase include the Franciscan terranes of California and Oregon, the Pacific Rim Complex of Vancouver Island, and the Chugach, McHugh, and Prince William terranes of southern Alaska. As is typical for Cordilleran events, accretion began earlier in the south (early Cretaceous) than it did to the North (late Cretaceous).

PHASE 6. NORTHWARD DISPERSAL OF TERRANE FRAGMENTS BY STRIKE–SLIP FAULTING

The final phase of development of the Cordillera has been dominated by strike–slip faulting and terrane fragmentation and dispersal in a complex transform system. Minor changes in plate motions have produced alternating periods of transtension and transpression, with resulting basin formation and basin compression that have continued throughout the Cenozoic. Current strike–slip activity along the San Andreas fault in California is just one manifestation of this type of tectonic activity that has effected the entire width of accreted crust throughout the entire Cordillera at one time or another.

Discussion

R. M. BURT (*University College of Wales, Cardiff, U.K.*). Dr Jones showed rocks interpreted as island or seamount in a large wide oceanic basin. These seamounts are now located inboard of a sequence identified as a volcanic arc.

Professor Dewey in his lecture included a slide of the Western Pacific. This area has abundant seamounts. He suggested that a compressional event closing the Pacific and emplacing the seamounts against a continent may lead to the defining of the seamounts as exotic terranes when in fact subduction zone parallel movement may not have been a factor in emplacement.

Does Dr Jones think that an initial westward-dipping subduction zone with seamount chains being located on the subducting plate followed by reversal of the subduction zone could not equally explain the above relationship? What are the boundaries of these terranes? Unless definite Palaeomagnetism or structural considerations prove that strike–slip subduction parallel mount these rocks can also be explained by conventional closure models and Deweygrams instead of strike–slip terrane-style movements.

It has been shown clearly along the Hawaii–Emperor chain that seamount chains young in one direction. Can the younging towards the north of this area of seamount be explained by this process. Would this younging also not imply straightforward closure, since terranes and strike–slip development result in a complex collage?

D. L. JONES. The longest chain of oceanic islands known in the Pacific Ocean (Hawaiian–Emperor chain) is about 4000 km long and spans about 70 my (= 57 km/1 my). Angayucham seamounts span at least 150 my but are of unknown length. If we assume they were generated above a single hot spot with average plate velocities similar to those that generated the H–E chain, then more than 8000 km of oceanic crust passed over the hot spot from late Devonian to early Jurassic time. More complex scenarios can be envisioned that might reduce the length of oceanic crust that passed over the hot spot, such as multiple hot spots or retrograde plate motion, but it is hard for me to escape an interpretation that the oceanic realm in which these seamounts formed was large and complex. In one place, Angayucham seamounts appear to young to the south, away from the continental margin on which they were obducted. But this relation has not yet been demonstrated on a regional basis, and indeed, nearby the youngest rocks are nearest the continent. Nor can it yet be shown that accretion of the seamounts occurred over a long time span, as Mesozoic rocks occur throughout the entire width of the accreted terrane. Island arcs of various and widely differing ages lie outboard of the obducted seamount-bearing terranes. Their palaeogeographic and genetic relations remain poorly understood and controversial. Despite this, it is clear that westward subduction beneath a single arc cannot explain the complexities in timing and style of emplacement of all the oceanic terranes that stretch along most of the Cordilleran margin, simply because no arc possesses the requisite characteristics in terms of timing of volcanic activity. A final point that favours transpressive emplacement of the oceanic terranes is that locally the bounding faults are clearly strike–slip in character.

Palaeomagnetic evidence bearing on the evolution of the Canadian Cordillera

By E. Irving, F.R.S., and P. J. Wynne

Pacific Geoscience Centre, Geological Survey of Canada, Box 6000, Sidney, British Columbia, Canada V8L 4B2

Palaeomagnetic data from Permian, Triassic and Jurassic bedded rocks, to which attitudinal corrections can be applied, yield palaeolatitudes concordant with those of ancestral North America, but very large predominantly anticlockwise rotations about vertical axes. Data from Cretaceous rocks yield apparent palaeolatitudinal displacements that increase westward. Small or negligible displacements are obtained from the Omineca Belt. Intermediate displacements (1000–2000 km) from the Intermontane Belt, are based on data from Cretaceous bedded sequences. Further to the west in the Coast Belt, larger apparent displacements (greater than 2000 km) have been obtained from plutons for which no attitudinal control is yet available. Data from Eocene rocks are concordant.

Possibilities to consider are as follows: (a) little or no displacement and tilting to the southwest at about 30°; (b) large (greater than 2000 km in the Coast Belt) northward displacement since mid-Cretaceous time preceded by southward displacement of comparable magnitude in Juro–Cretaceous time; (c) lesser (1000–2000 km) overall displacement coupled with variable and lesser tilts to the south and southeast of plutons of the Coast Belt. Under hypothesis (a) the western Cordillera was formed and has remained in approximately its present position relative to ancestral North America; data from bedded volcanics of the Intermontane Belt are not consistent with this hypothesis. From the evidence currently available we favour hypotheses (b) or (c), although more data from bedded sequences are required. It is noteworthy that hypotheses (a) and (c) predict tilt directions that differ by about 90° and hence ought to be distinguishable by geological studies.

1. Introduction

Our purpose is to review the palaeomagnetic evidence that bears on the evolution of the Pacific Northwest sector of the North American Cordillera. We are concerned mainly with data from British Columbia, Yukon and adjacent areas of Alaska and Washington, but we review also Cretaceous data from California and Baja California.

Palaeopoles calculated from the directions of magnetization (palaeodirections) observed in rock-units laid down contemporaneously on the craton of North America and in the Cordillera should agree if both have formed in their present relative positions. If they do not, and if post-depositional tilting of the rock-units can be estimated, then the palaeolatitudinal displacement and rotation relative to the craton of the Cordilleran locality can be calculated. The accuracy of determinations depends on the accuracies with which cordilleran and cratonic reference palaeopoles can be positioned and dated. The errors are such that displacements less than about 500 km and rotations less than 5° are unlikely to be identified. Relative palaeolongitudinal displacements cannot be determined by this method.

31

Terranes and belts

The Western Cordillera in British Columbia and adjacent areas of northern Washington and Alaska comprise several morpho-geologic belts which trend general NNW to SSE (figure 1). Within each belt are several fault-bounded terranes (figure 2). Some terranes are confined within a single belt, others are not. For example, the Wrangellia terrane straddles the boundary between the Intermontane and Coast belts, and Quesnellia occurs on both sides of the boundary between the Intermontane and Omineca belts (compare figures 1 and 2).

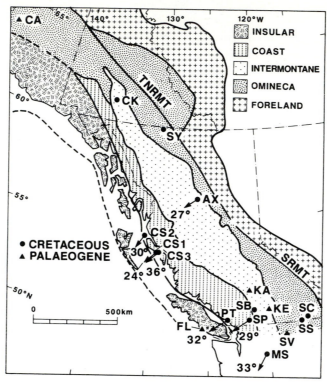

FIGURE 1. Morpho-geological belts showing sampling localities in Cretaceous and Lower Tertiary rocks. Arrows at localities of Cretaceous plutons in the Coast and Intermontane belts indicate the direction and magnitude of tilting required to explain their anomalous palaeodirections. Localities labelled as in table 2. NRMT and SRMT are the northern and southern sections of the Rocky Mountain trench.

The concept of terranes, as it applies to the Canadian Cordillera, relates to the formation and subsequent amalgamation of geologically distinct and tectonically separate rock assemblages, processes that happened mainly during Triassic through early Cretaceous time (Monger *et al.* 1982). Belts, on the other hand, are defined largely on the basis of physiography, and have been produced by the accretion of terranes to North America and their subsequent deformation.

Although the accretion of some young westerly terranes (the Crescent Terrane for example) occurred in the Eocene and overlapped in time the processes that created the morpho-geologic belts, the distinction between an earlier period of terrane formation and amalgamation, and a later period during which belts were established is valid for most of the Canadian sector of the Cordillera. Hence, we consider palaeomagnetic data from older rock-units (Permian through Lower Jurassic) in the terrane context, and data from younger rock-units (mid-Cretaceous through Eocene) in the context of the distribution of belts.

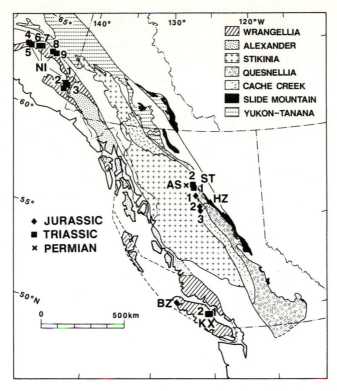

FIGURE 2. Major terranes with sampling localities in Permian, Triassic and Jurassic rocks.

Themes

Discussions of Cordilleran palaeomagnetism have centred about four main themes, beginning with the discovery of large rotations about vertical axes. These were first observed as aberrant declinations in bedded Eocene volcanic rocks (Cox 1957), although not at the time recognized as indicative of tectonic rotations (Cox 1980). Subsequent work has shown rotations to be common, typically clockwise in Cretaceous and Palaeogene rocks, and often anticlockwise in older sequences (Beck 1976, 1980; Monger & Irving 1980; Irving & Yole 1987).

The second theme concerns the apparent displacement from the south of some elements of the Cordillera by as much as 2000 km since mid-Cretaceous time. The idea that such late displacements could have occurred emerged from the work of Beck & Noson (1972) on the Mount Stuart batholith of the northern Cascade Mountains of Washington (MS of figure 1), and of Tiessere & Beck (1973) on the Peninsular Batholith of southern California (SA of figure 5). Both intrusions are approximately 100 Ma old, and both yielded anomalously low inclinations (i.e. low palaeolatitudes). Several years later similar discordant magnetizations were observed in the Cretaceous Axelgold intrusion, a body which is located in the eastern part of the Intermontane Belt of British Columbia (Monger & Irving 1980). This and other evidence led Monger & Irving (1980) and Irving *et al.* (1980) to propose that about two-thirds of British Columbia and adjacent parts of Washington and Alaska (a region which later was referred to as *Baja British Columbia* (Irving 1985)) had been displaced from the south in latest Cretaceous and Palaeocene time. The concept of Baja British Columbia, (or Baja B.C.) was developed from the work of Packer & Stone (1974), and from Atwater's study of the northward motion of Baja California (Atwater 1970). However, the term 'Baja B.C.' is meant to imply only that the region formerly had a lower or more southerly position, and that the motions which brought it into its present position was predominantly in a coastwise sense, like that of Baja California today, but no specific mechanism of transport is implied.

The third theme concerns data from older sequences (Permian, Triassic, Jurassic) which have yielded apparent displacements that are very much less than those obtained from Cretaceous rocks. Initially, these data indicated motions from the south in excess of 1000 km (Symons 1971b; Irving & Yole 1972, 1980; Hillhouse 1977), but revisions of the timescale (for example by Harland *et al.* 1982) have forced a recalibration of the apparent polar wander path for cratonic North America, causing a decrease in estimates of displacement to the point that many of them may not be significant (Gordon *et al.* 1984; May & Butler 1986). These discoveries of large net displacements from younger rocks and small net displacements from older rocks has led to two divergent discussions in the palaeomagnetic literature. In one, workers have entertained the possibility that the terranes of the Pacific Northwest sector of the Cordillera first moved south *en bloc* by 1000 km or more, and then north by as much as 2000 km (Irving *et al.* 1985; Beck 1990; Irving & Wynne 1990). In the second discussion, the small or negligible net displacements estimated from older rocks is deemed the more important discovery, and the larger displacements determined from younger rocks are explained by appealing to systematic tilt, thus denying large latitudinal displacements (May & Butler 1986; Butler *et al.* 1989) From the latter has emerged a conservative 'fixist' view of the tectonics of the Cordillera by which terranes of the Pacific Northwest are believed to have formed essentially where they are now, and to have moved no more than small distances in a latitudinal sense.

The fourth theme concerns the attenuation and dispersal of crustal elements in the Cordillera. This was first noted in Wrangellia, when localities now 1500 km apart yielded similar palaeolatitudes (Hillhouse 1977; Yole & Irving 1980) indicating that localities, once close together, have since been pulled apart. Other instances have been discussed for tectonic units situated inboard of Wrangellia (Irving *et al.* 1985; Irving & Monger 1987; Umhoefer 1987).

Some difficulties

Although the principle is the same as that used in the 1950s to test Wegener's hypothesis of continental drift, the practice is more difficult for several reasons. Firstly, the apparent relative displacements are 2–5 times less than those among continents so that some studies have been made at the limits of resolution of the method. Secondly, in orogenic regions it is difficult to obtain accurate, well-dated records of the palaeofield, because thermal, structural and diagenetic histories of rocks which control their magnetization, are complex and only partly understood; magnetic overprinting often occurs at times that are ill defined relative to deformation, making the separation of pre-, syn- and post-tilting magnetizations an ever-present and sometimes insoluble problem. The third difficulty concerns the interpretation of data from plutonic rocks in orogenic belts. In stable cratons, plutons generally have not undergone post-emplacement tilting and serve as excellent recorders of the palaeofield, but in orogenic zones it is generally difficult to determine post-emplacement tilt, so the interpretation of palaeomagnetic data from them can be ambiguous. Hence, discordant data from plutons indicate either displacement or post-emplacement tilt, or some combination of the two (Beck *et al.* 1981). However, sufficient detailed studies have now been made to allow these problems to be addressed although not yet settled.

TABLE 1. CRATON REFERENCE PALAEOPOLES

	T (Ma) craton	T (Ma) cordillera	lat.° N, long.° E (A_{95})	rel.	
A	Eocene	54–48	55–48	83, 176 (3)	H	KE, KA, SP, FL
B	Palaeocene	67–62	Palaeocene (66–58)	81, 185 (6)	L	CA
C	latest Cretaceous and Palaeocene	73–63	70	78, 186 (8)	L	CK
D	mid-Cretaceous	136–85	120–95	71, 196 (5)	H	SC, SS, SA, AX, SB, CK, CS, SP, PP, MS
E	early Jurassic	195–191	Sinemurian–Toarcian (204–187)	65, 082 (4)	M	HZ, BZ
F	late Triassic	Norian–Carnian (230–208)	late Ladinian–early Norian (332–323)	53, 098 (7)	M	ST, KX, NI
G	early Permian	285–255	late Sakmarian–early Artinskian (270–265)	46, 119 (4)	H	AS

Notes for table 1. First column gives ages for which data are available from both craton and cordillera. Age spans of respective data are given in the second and third columns, in numbers if determined radiometrically, and by geological stages if determined stratigraphically followed by numerical ages estimated from the timescale of Palmer (1983). Palaeopoles and errors ($P = 0.05$) are followed by an estimate of reliability (H, high; M, moderate; L, low). The last column gives the cordilleran rock-units for which displacements and rotations have been calculated in table 2. The derivation of palaeopoles E and F is detailed below. A, B, C, D, and G have been given elsewhere, A, Irving & Brandon (1990); B, Jacobsen et al. (1980); C, Marquis & Globerman (1988); D, Globerman & Irving (1988); E, derived from following Newark Supergroup rocks that have been unaffected by Middle Jurassic overprints: North Mountain basalts, Nova Scotia, 191 Ma, 66.4° N, 71.9° E, $A_{95} = 11°$ (Hodych & Hayatsu 1988); Connecticut Valley rocks, 193 Ma, 65° N, 87° E, $A_{95} = 10°$ (de Boer & Snider 1979); prefolding magnetization of intrusive rocks Connecticut and Maryland, 91 Ma, 63.1° N, 82.5° E, $A_{95} = 2.8°$ (Smith & Noltimier 1979); Piedmont dykes, 195 Ma, 66.1° N, 83.9° E, $A_{95} = 7.5°$ (Dooley & Smith 1982); F, is the mean of following three data: Middle Carnian to early Norian rocks (Stockton, Lochatong and Passiac formations) of Newark Supergroup 53.6° N, 101.6° E, $A_{95} = 4.8°$ (Witte & Kent 1989); Popo Aige Formation, Norian and Carnian of Wyoming 55.5° N, 95.5° E, $A_{95} = 8°$ (Grubbs & van der Voo 1976); the Abbott (228 Ma) and Agamenticus (221 Ma) plutons of Maine, mean of 16 sites in Wu & van der Voo (1988) is 188.7°, −4.9°, $k = 241$, $\alpha_{95} = 2.4°$ with palaeopole 48.3° N, 96.0° E, $A_{95} = 1.9°$; G, mean of 9 palaeopoles in Irving & Irving (1982).

Table 2. Cordilleran data

rock-unit	ref.	lat.° N, long.° W	D°, I°	α95	lat.° N, long.° E	A95	λp°	RPD°	RR°
Palaeogene									
KE, Kelowna volcanics (52 Ma) IMB	pc	49.9, 119.7	352, 69	6	85, 197	10	53±08	00±08	-02±14
KA, Kamloops Group, volcanics (49 Ma) IMB	tc	51.0, 121.3	355, 73	7	81, 222	12	59±10	-05±10	-06±19
SV, Sanpoil volcanics (53–48 Ma) Wa., IMB	tc	48.5, 118.5	016, 69	4	79, 305	6	52±04	-01±05	-26±08
FL, Flores volcanics (51 Ma) INB	tc	49.0, 125.5	350, 69	7	81, 188	10	53±08	00±08	00±14
CA, Cantwell Formation, volcanics (61 Ma) Alaska	tc	63.6, 148.8	143, −85	5	70, 195	11	81±10	-09±09	—
Cretaceous									
Pacific Northwest (British Columbia, Yukon Territory and Washington)									
SC, Skelly Creek batholith (95 Ma) OB	nc	49.3, 116.5	349, 74	6	78, 216	10	60±08	-01±08	-17±17
SS, Summit Stock (104 Ma) OB	tc	49.1, 117.1	318, 74	3	64, 194	5	60±04	-01±06	14±11
SY, Permian limestone, overprint, OB	nc	59.4, 129.9	328, 77	4	76, 172	9	65±06	08±07	-05±20
AX1, Axelgold intrusion, gabbro (120–90 Ma) IMB	nc	56.2, 126.1	032, 60	4	64, 348	6	41±05	27±06	-65±12
AX2, Axelgold intrusion, gabbro (120–90 Ma) IMB	tc	56.2, 126.1	025, 69	5	76, 327	8	53±06	16±07	-58±15
SB, Spences Bridge Group, volcanics (104 Ma) IMB	tc	49.9, 121.0	039, 64	5	64, 321	7	46±06	16±07	-66±12
CK, Carmacks Group (70 Ma), volcanics Yukon, IMB	tc	61.2, 135.5	167, −71	5	82, 109	8	55±06	14±09	-09±21
CS1, Captains Cove pluton (109 Ma) CB	nc	53.2, 129.4	021, 61	7	72, 351	10	42±08	25±09	-49±14
CS2, Stephens Island pluton (102 Ma) CB	nc	54.1, 130.6	013, 53	5	67, 021	9	34±07	34±08	-41±13
CS3, Gil Island pluton (136 Ma) CB	nc	53.2, 129.2	029, 50	17	59, 356	19	31±15	36±16	-57±21
CS, Three coast pluton combined, CB	nc	53.5, 129.7	022, 56	6	67, 360	8	37±06	30±07	-50±13
SP, Spuzzum pluton (104 Ma) CB	nc	49.4, 121.5	031, 57	6	65, 017	7	38±06	23±07	-58±11
PT, Porteau pluton (100 Ma) CB	nc	49.5, 123.2	031, 54	9	63, 345	11	35±09	27±09	-58±13
MS, Mount Stuart batholith (100 Ma) Wa. CB	nc	47.5, 121.0	010, 45	5	68, 035	5	27±04	32±06	-36±09
Klamath Mountains–Sierra Nevada									
KM, Great Valley sequence, Ca. sst.	tc	40.3, 122.7	338, 65	6	73, 177	8	47±07	06±09	01±14
SN, Sierra Nevada plutons, Ca.	nc	37.5, 119.0	—		72, 175	6	43±04	06±06	03±09
MD, Mina deflection, plutons, Nv.	nc	38.5, 118.3	356, 64	11	81, 226	16	47±15	03±15	-18±23
BV, Bear Valley Springs pluton, Ca.	nc	35.2, 118.3	022, 56	6	72, 322	7	37±06	10±08	-42±10
Baja California (Mexico and U.S.A.)									
SA, Ladd and Williams Fms seds, Ca.	tc	34.5, 117.0	324, 45	7	58, 149	7	27±06	19±07	16±09
AL, Alisitos Fm, Baja Ca.	tc	31.5, 116.5	010, 50	8	81, 336	9	31±08	12±09	-29±11
LB, La Bocana Roja Fm, Baja Ca.	tc	30.0, 116.0	012, 52	6	80, 317	6	33±05	-09±07	10±09
PR, Peninsular Ranges batholith, Ca. and Baja Ca.	nc	30.0, 115.0	006, 48	3	85, 346	3	29±03	12±06	-25±07
VF, Valle Fm, turbidites, Baja Ca.	tc	27.5, 114.5	006, 40	4	83, 013	4	23±03	16±06	-25±07
ER, El Rosario Formation, Baja Ca.	tc	31.3, 116.3	005, 45	3	83, 021	2	26±01	16±04	-24±06
Western California									
LL, Laytonville Limestone age, Albian–Cenomanian	tc	39.5, 123.6	—		—	—	-14±05	66±07	—
PP, Pigeon Point Fm, Upper Cretaceous turbidites	tc	37.2, 122.5	, 37	5	—	—	21±04	29±06	—
CL, Calera Limestone, mid-Cretaceous	tc	37.5, 122.3	—		—	—	24±04	27±06	—
FT, Figueroa Mt, Upper Cretaceous turbidites	tc	35.0, 120.0	023, 12	11	54, 019	8	06±06	41±08	-43±09
JA, Jalama Fm, Upper Cretaceous turbidites	tc	34.0, 120.4	024, 44	3	68, 345	3	26±02	11±05	-01±07
MC, Marin County basalt, Valinginian	tc	38.0, 122.8	078, 47	8	26, 309	9	28±07	23±10	98±13

Jurassic (British Columbia and Washington)

Unit	Corr.	Location	$D°, I°$	α_{95}	Palaeopole	A_{95}	λ_p	RPD	RR
HZ1, Hazelton Group, Jl (ST)	tc	56.5, 126.8	114, −52	25	40, 144	30	33±24	01±25	52±29
HZ2, Hazelton Group, Jl (ST)	tc	55.8, 126.6	242, 56	19	17, 185	22	37±18	−04±19	105±22
HZ3, Hazelton Group, Jl (ST)	tc	55.6, 126.4	359, 55	16	70, 057	19	36±16	−03±17	−12±19
HZ, Hazelton Group, volcanics, Jl (ST)	tc	56.0, 126.6	—, 54	—	—	—	35±16	−01±16	—
BV, Bonanza Group, volcanics, Jl (WT)	tc	50.5, 128.1	276, 42	6	22, 154	8	25±06	04±07	71±07
JI, James Island Fm, turbidites, Ju, Wa. (DT)	tc	48.5, 122.8	344, −01	18	39, 078	14	01±01	37±12	−19±12

Triassic (British Columbia and Alaska)

Unit	Corr.	Location	$D°, I°$	α_{95}	Palaeopole	A_{95}	λ_p	RPD	RR
ST1, Stuhinni Group, Tru (ST)	tc	56.7, 126.4	300, 44	6	38, 133	10	26±08	−01±09	32±10
ST2, Stuhinni Group, Tru (ST)	tc	56.6, 126.5	281, 38	7	24, 146	9	21±07	03±09	51±09
ST, Stuhinni Group, volcanics combined, Tru (ST)	tc	56.7, 126.5	—, 41	—	—	—	23±07	02±09	—
KX1, Karmutsen Formation, lavas, sills, Tru (WT)	tc	49.9, 125.6	337, −24	9	24, 080	8	13±06	06±08	177±08
KX2, Karmutsen Formation, Tru (WT)	tc	49.9, 125.6	013, −34	5	20, 042	5	19±04	00±06	141±07
KX, Karmutsen Formation combined, Tru (WT)	tc	49.9, 125.6	003, −33	6	23, 052	6	18±05	01±07	−29±07
NIA, Nicolai volcanics 1–3 combined, Ak., Tru (WT)	tc	61.6, 142.3	074, −20	6	02, 146	5	10±03	23±08	−133±09
NI1, Nicolai volcanics, Ak. (WT)	tc	61.7, 142.4	077, −23	3	−05, 325	2	12±02	22±05	64±06
NI2, Nicolai volcanics, Ak. (WT)	tc	61.6, 142.9	065, −18	8	03, 333	7	09±02	25±07	105±08
NI3, Nicolai volcanics, Ak. (WT)	tc	61.5, 142.8	095, −15	17	−10, 306	14	08±11	26±12	45±13
NIB, Nicolai volcanics, 4–9 combined, Ak. (WT)	tc		—	—	—	—	14±04	22±07	—
NI4, Nicolai volcanics, Ak. (WT)	tc	63.1, 147.4	060, −33	9	−04, 338	9	18±08	18±08	103±10
NI5, Nicolai volcanics, Ak. (WT)	tc	63.1, 147.1	138, −32	11	−37, 265	8	18±06	19±08	01±09
NI6, Nicolai volcanics, Ak. (WT)	tc	63.2, 146.3	084, −20	12	−07, 313	8	10±06	26±08	53±09
NI7, Nicolai volcanics, Ak. (WT)	tc	63.3, 145.9	095, −07	7	−05, 302	6	03±05	33±07	43±08
NI8, Nicolai volcanics, Ak. (WT)	tc	63.1, 144.4	105, −30	13	−22, 300	10	16±08	19±10	34±11
NI9, Nicolai volcanics, Ak. (WT)	tc	63.1, 144.3	119, −27	25	−30, 288	24	15±19	21±20	20±21

Permian (British Columbia)

Unit	Corr.	Location	$D°, I°$	α_{95}	Palaeopole	A_{95}	λ_p	RPD	RR
AS, Asitka Group, basalts, Pl (ST)	tc	56.7, 126.6	129, −40	8	40, 123	5	23±04	04±05	06±06

Notes for table 2. First column gives the rock-unit name, age, province or state of origin (those undesignated are from British Columbia), and, when in Canada, the tectonic belt or terrane (bracketed) as follows. IMB Intermontane, OB Omineca, and CB Coast belts; ST Stikine, DT Decatur and WT Wrangel terranes. Second column gives the attitudinal corrections applied, tc total correction, pc partial correction, nc no corrections. Column three gives the sampling location. $D°, I°$ are the declination and inclination of the mean direction of remanent magnetization, α_{95} is the radius of the circle of confidence ($P = 0.05$). Column six contains the palaeopole and A_{95} is its radius of the circle of confidence. λ_p is the palaeolatitude of the sampling locality, RPD is the relative palaeolatitudinal displacement (positive if the motion is northward), and RR the rotation (negative if clockwise), both relative to the reference palaeopoles of table 1 for the northern option. Values for southern option are not listed, but can be readily calculated from table 1. The references are as follows: KE, Bardoux & Irving (1989); KA, Symons & Welling (1989); SP, Fox & Beck (1985); FL, Irving & Brandon (1990); CA, Hillhouse & Grommé (1990); SC and SS, Irving & Archibald (1990); SY, Butler et al. (1988); AX, Monger & Irving (1980); Armstrong et al. (1985); CS, Symons (1977); SB, Irving & Thorkelson (1990); SP, Irving et al. (1985); PT, Porteau pluton based on high unblocking temperature (755 °C) high coercive force (greater than 80 mT) magnetizations from five sites which fulfill the most stringent (AI) criteria in Irving et al. (1985), data of Irving et al. (1985) and unpublished data of Irving & Yorath; TS, Beck (1975); MS, Beck et al. (1981); KM, Mankinen & Irwin (1982); JM, Russell et al. (1982); SN, Frei et al. (1984), Grommé & Merrill (1965), Frei (1986) mean of four plutons; MD, Geissman et al. (1984); BV, Kanter & McWilliams (1982); SA, Fry et al. (1985); AL, LB, and VF, Hagstrum et al. (1985); PR, Teissere & Beck (1973), Hagstrum et al. (1985); ER, Filmer & Kirschvink (1989); LL, Alvarez et al. (1980), Tarduno et al. (1986); PP, Champion et al. (1984); CL, Courtillot et al. (1985); Tarduno et al. (1985); FT, McWilliams & Howell (1982); JA, Champion et al. (1986); MC, Grommé (1984); HZ, Monger & Irving (1980); BV, Irving & Yole (1987); ST, formerly Takla Group, Monger & Irving (1980); KX, KX1 NW group, KX2 NNE group calculated from Yole & Irving (1980) and Schwarz et al. (1980) sites unit weight; NIA McCarthy Quadrangle, Hillhouse (1977) three localities; NIB Mt, Hayes & Healey Quadrangle, Hillhouse & Grommé (1984) six localities; AS, Irving & Monger (1987).

We consider only data from rocks that are well dated, and for which there are good reasons to assume that they record the palaeofield at, or soon after, deposition. With one exception (SY of table 2), we do not consider magnetizations that could be overprints because they generally lack an adequate time basis for tectonic discussions. Finally, only data from rock-units that may be expected to have averaged out palaeosecular variation are included. Data that has not been accepted are listed in the appendix. Data available in abstract are noted, but are not integrated into the analysis.

2. Cratonic reference palaeopoles

Data from the Pacific Northwest fall into four main (Eocene, mid-Cretaceous, early Jurassic, late Triassic) and three subordinate (Palaeocene, latest Cretaceous, Permian) groups. Reference palaeopoles have been obtained by selecting data from the craton which span each of these seven time intervals (table 1). In this way errors in time correlation between craton and cordillera are minimized. The rationale for using this, rather than other procedures (see, for example, Gordon *et al.* 1984) has been given by Irving & Yole (1987). Only the reference palaeopoles for the Eocene, mid-Cretaceous and early Permian can be regarded as of high reliability. Others are based on fewer, often less well-based data.

3. Tertiary

Many studies of Neogene rock-units have yield palaeopoles that agree with the present geographic pole and with palaeopoles from the craton (reviewed Irving & Wynne 1990). Three studies of Eocene rocks from British Columbia (FL, KA, KE) yielded results that are concordant with the craton (figure 3). Data from eastern Washington (SV) gave a concordant palaeolatitude, but a clockwise rotation. Data from the Cantwell Formation (CA) give very high palaeolatitudes with a mean displacement of 9°, which is marginally significant. Note, however, that the reference palaeopole for the Palaeocene is based on only a single

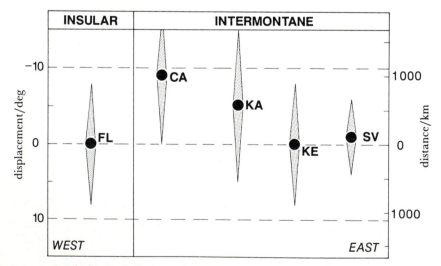

FIGURE 3. Apparent palaeolatitudinal displacement relative to ancestral North America estimates from Palaeocene and Eocene rocks. Labelled as in table 2 and figure 1. Recent data from Eocene rocks of central Intermontane Belt similarly yield no palaeolatitudinal displacement (Vandall & Palmer 1988).

determination (table 1). In these and other analyses of table 2 and figures 4, 7 and 8, 95% errors are given, so that if departures are found they must be regarded as statistically significant. It is important to note that these errors are not ranges but are probability distributions. The probability is highest at, or close to, the mean and diminishes away from it. For this reason error bars are shown not as lines of equal thickness, but as outwardly directed arrow heads signifying the fall in probability with distance from the mean. The data of figure 3 show that the major elements of the Pacific Northwest sector of the Cordillera were essentially in place by 50 Ma.

4. CRETACEOUS DATA FROM THE PACIFIC NORTHWEST

Apparent displacements estimated from Cretaceous rocks of British Columbia and Washington are shown in figure 4. Data are grouped according to the degree of attitudinal control. Apparent displacements increase from east to west, and rotations are predominantly clockwise (figure 5).

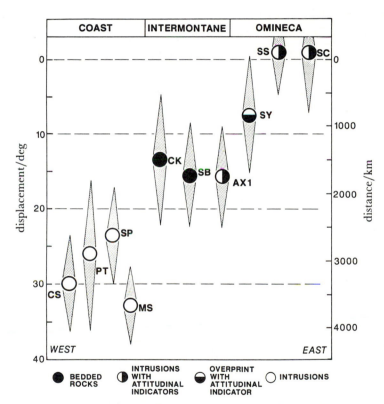

FIGURE 4. Apparent palaeolatitudinal displacements estimated from Cretaceous rocks. Labelled as in table 2 and figure 1.

In the south, data from the eastern Omineca Belt, not far to the west of the Southern Rocky Mountain trench (SS and SC of figure 1), show no latitudinal displacement. These data are from plutons, but corrections for tilts have been made using bathozonal information (Irving & Archibald 1990). Evidently this region has remained nearly fixed to the margin of ancestral North America since mid-Cretaceous time.

In the north, the Slide Mountain terrane, which is situated west of the Northern Rocky

Mountain Trench fault, has yielded an apparent displacement of 800 ± 850 km (SY, figure 1 and table 2). This is marginally significant, but is, nevertheless, consistent with geologically based estimates of displacement of at least 750 km on the Tintina and northern Rocky Mountain Trench and associated faults, which are located to the east of the sampling locality (Gabrielse 1985). SY is derived from an overprint observed in Permian limestones of the Slide Mountain Terrane. This magnetization is presumed to be contemporaneous with that of the nearby mid-Cretaceous Cassiar batholith, and the sub-horizontal thrust that underlies the sampling area indicates that post-intrusion tilting has been negligible. This datum is, therefore, accepted as a reasonable indicator of modest displacement.

Displacements estimated from the Spences Bridge and Carmacks Groups, both stratified volcanic units in the Intermontane Belt, are significant, and are in mutual agreement. It should be noted, however, that the Carmacks Group was laid down in the latest Cretaceous (70 Ma) at a time when the cratonic reference field is not well known and may have been changing rapidly (table 1). Both studies include data from beds which yield a positive tilt test, indicating that the magnetization was acquired before deformation. The Spences Bridge has rotated clockwise. The young Carmacks has not undergone significant rotation. Displacements obtained from bedded sequences also are in good agreement with that estimated for the Axelgold intrusion after correction to its prominent planes of layering (Monger & Irving 1980). If Axelgold magnetizations are considered to have been acquired after the layers were tilted, then the estimated displacement approaches 3000 km (table 2).

All data from the Coast Belt are from plutonic rocks. Mount Stuart, in the northern Cascade Mountains of Washington, has the largest apparent displacement, exceeding 3500 km. Alternatively, it may have been tilted down by about 33 at 210° E (figure 1). Three coastal plutons in the Prince Rupert area, detailed separately in figure 1 and table 2, have a mean apparent displacements of over 3000 km, or a mean apparent tilt of down 28° at 223°. The Spuzzum and Porteau plutons in the southern Coast Ranges give displacements of about 2600 km, or mean apparent downward tilts of 29° at 238° E and 32° at 240° E respectively. All apparent displacements exceed those observed from the Intermontane Belt, and all apparent rotations are about 60° clockwise (table 2 and figure 5). The question is, have the plutons been tilted or have they been displaced latitudinally?

Irving & Thorkelson (1990) have attempted an answer by comparing data from the bedded volcanics of the Spences Bridge Group of the Intermontane Belt, and the nearby Spuzzum pluton of the Coast Belt. The apparent displacement of the former is just significantly less than that of the latter (figure 4). The two mid-Cretaceous units are separated by a series of dextral strike–slip faults, which delineate the belt boundary, and which were active in late Cretaceous and early Tertiary times. Monger (1990) and Umhoefer et al. (1989a) have estimated a total displacement of between 300 and 400 km along them. Adding 350 km to the displacement of 1750 ± 800 km estimated for the Spences Bridge Group, produces a total expected displacement of the Spuzzum pluton of about 2100 ± 800 km. This is not significantly different from the displacement estimated from the untilted pluton itself (2600 ± 700 km, table 2). Hence, it appears that most but not necessarily all of the apparent displacement observed from the Spuzzum pluton is a real latitudinal offset, and is unlikely to have been caused entirely by 30° tilting to the southwest as depicted in figure 1.

Displacements estimated for the Prince Rupert and Mount Stuart intrusions are larger (about 3000 km). Hence, the total offset elsewhere on faults between the Coast and

40

Intermontane belts either has been larger than estimated, or these intrusions have been tilted relative to Spuzzum as well as translated. The tilting about a horizontal axis required to bring the palaeodirection from the Coast (CS) plutons into accord with that from Spuzzum is $6° \pm 6°$ at $157°$, that for the Porteau Pluton is negligible, and that for Mount Stuart is $18° \pm 7°$ at $142°$.

Alternatively one may argue that all differences between apparent displacements from the Intermontane and Coast belts are caused by tilting generally to the south or southeast. The tilts required to bring data from the plutons of the Coast Belt into agreement with data from bedded volcanics of the Spences Bridge Group in the Intermontane Belt are as follows: coastal plutons (CS) down $12°$ at 168, Porteau plutons (PT) $8°$ at $194°$, Spuzzum pluton (SP) $08°$ at $176°$, Mount Stuart batholith (MS) down $25°$ at $154°$.

We may summarize now the three possible explanations of the Cretaceous data of figure 4: (a) regional tilting to the west and southwest by about $30°$ (figure 1); (b) real displacements increasing westward to over 2000 km; (c) real displacement of about 1500 km, the additional aberrancies of the Coast Belt being caused by variable tilt of plutons to the south and southeast. Data presently available from bedded rocks of the Spences Bridge group are inconsistent with (a). It should be possible to test the validity of (a) or (c) using geological studies because the required tilts are in directions approximately $90°$ apart.

If hypothesis (b) is correct, then there must have been not one but many fault systems along which transcurrent motion occurred. Moreover, the clockwise rotations commonly observed (table 2 and figure 5) show that some of the apparent strain may have been accommodated by block rotation. Therefore, to make quantitative geological tests of palaeomagnetically estimated displacements, it is necessary to determine displacements on all late Cretaceous and Palaeocene faults and all the associated block rotations that have occurred to the west of the Rocky Mountain Trench in the north, and west of the boundary between the Omineca and Intermontane belts in the south. It is not a sufficient test to analyse a few fault systems, as Price & Carmichael (1986) have done. It is noteworthy that where such comparisons have been made in juxtaposition to palaeomagnetically estimated displacements, notably along the Northern Rocky Mountain and the Fraser–Yalakom fault systems, the displacements obtained are broadly consistent with the palaeomagnetic data, as described above.

4. Cretaceous results, California and Baja California

It should be possible to integrate the data of figure 4 with that from mid-Cretaceous rocks elsewhere in the Cordillera, and an attempt at integration might help to distinguish among the three possibilities identified above. We now attempt to do this by reviewing data from California and Baja California (table 2). For space reasons, Cretaceous data from terranes north and west of Wrangellia are not considered.

In figure 5, the data are arranged in four regional groups, the Pacific Northwest (just described), the Klamath Mountains and Sierra Nevada, Baja California, comprising Baja California and the adajacent Peninsular Ranges batholith of southern California, and Western California, comprising California west of the Great Valley sequence. Palaeolatitudes are plotted against the modern west coast of North America whch runs approximately north–south on the mid-Cretaceous cratonic reference grid. Rotations are shown in insets. All magnetizations, like those further north, have positive inclinations and can be ascribed to the

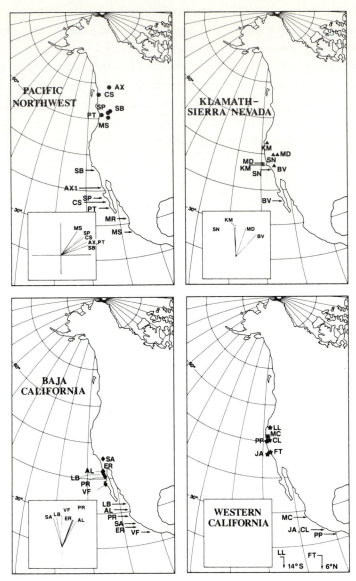

FIGURE 5. Mid-Cretaceous rocks of the Cordillera grouped by region of derivation. Alaskan data not included. Dots (Pacific Northwest), triangles (Klamath–Sierra Nevada), diamonds (Baja California) and stars (Western California) are sampling localities labelled as in table 2. Arrows are the corresponding estimated palaeolatitudes. The grid is drawn relative to the mid-Cretaceous cratonic palaeopole of table 1. Rotations relative to cratonic North America are shown in insets, the convention being that if rotation were zero the line would be along the vertical axis.

Cretaceous Normal Superchron (Harland *et al.* 1982). Consequently, all regions, except possibly parts of Western California, were in northern palaeolatitudes in the Cretaceous. Palaeolongitude is indeterminate, so localities marked in figure 5 could have lain anywhere to the west within the palaeolatitudinal belts indicated.

Klamath Mountains–Sierra Nevada

One datum (KM) is from bedded rocks and attitude corrections have been applied. Three data are from intrusive rocks and no corrections have not been made. Except for BV, which is from the southern end of the Sierra Nevada, they all are in good agreement, and all displacements are small (table 2, figure 5). Rotations are negligible or small and clockwise. BV shows the largest displacement, although barely significant at the 95% confidence level; Kanter & McWilliams (1982) ascribe the 45° apparent rotation to bending of the southern end of the Sierra Nevada in response to dextral shear along the proto-San Andreas transform. Tilting to the southwest also could have been responsible for part of the divergence of BV.

Baja California

All apparent displacements of about 15° are from the south and are in good agreement. Rotations are variable. Of the six determinations, five are from bedded sequences and attitude corrections have been applied. The sixth, is based on extensive studies of the Peninsular Range batholith.

Western California

Data from the Franciscan Complex and 'Salinia' block of western California yield very large and generally variable displacements from the south. One, from the Laytonville Limestone (LL of figure 5 and table 2), in which age, top-direction, and bedding attitudes are well known, and whose magnetization has been subject to detailed study, implies a displacement exceeding 60°. Northward displacement of about 24° is inferred from studies of the Calera Limestone (CL) which, like the Laytonville Limestone, is a knocker in the Franciscan Complex. Apparently the Laytonville Limestone was deposited in the Southern Hemisphere, and the Calera in the Northern Hemisphere. According to the palaeomagnetic evidence the Franciscan Complex and Western California generally, is unlike the other three regions, and, contains rocks that have come from very different places. Some rock-units are from Cretaceous oceanic assemblages (e.g. Laytonville Limestone, Pigeon Point Formation) that may have been transported on oceanic plates from considerable distances. It is noteworthy that very recently data has been obtained from the Jurassic James Island Formation of the Deatur Terrane of the San Juan Islands, which indicate net displacement from the south in excess of 20° (Bogue *et al.* 1989). Although situated much further north at present, the Decatur Terrane (the San Juan Islands are just southeast of Vancouver Island) is an oceanic assemblage with many resemblances to the Franciscan Complex.

5. Reconstructing the mid-Cretaceous Cordillera

A major uncertainty in the interpretation of data from the western Cordillera is in the determination of the horizontal plane at the time of remanence acquisition (Beck *et al.* 1981; Irving *et al.* 1985). In the Sierra Nevada and Klamath region and Baja California, there is good agreement among data from intrusive and bedded sedimentary rocks (table 2, figure 5). In the Pacific Northwest, when due account is taken of transcurrent motions along faults separating palaeomagnetically studied bedded rocks and intrusions, there is also a fair measure of agreement. It could be argued that these agreements are fortuitous. For example the plutons could have been tilted and the inclinations in the bedded sedimentary sequences could have been flattened at deposition (the inclination error of King (1955)) or by later compaction, yielding palaeolatitudes that are too low in both rock types. This is unlikely in the Pacific Northwest where the bedded sequences studied are for the most part massive lava flows

(Marquis & Globerman 1988; Irving & Thorkelson 1990). Hence, the agreements between different rock-types, constitute good evidence that the estimates of palaeolatitude are correct within the errors stated.

Data from western California indicate large and internally variable displacements. In contrast, data from the other three regions of the western Cordillera yield post-mid-Cretaceous displacements that differ from region to region, but within each region they are generally in good internal agreement (figure 5). Cretaceous intrusions from these three regions are subduction related, and the sedimentary and volcanigenic rocks are all of shallow-water or terrestrial origin. It is unlikely, therefore, that these assemblages were formed far from the margin of cratonic North America. Hence it would seem that the terranes into which they were intruded or upon which they were deposited, had by then been accreted to North America, but not, according to the evidence of figure 5, in their present relative positions. It is as if the three regions were not formed in their present order. Instead, during the early Cretaceous, the Pacific Northwest was interposed between the Klamath Mountains–Sierra Nevada and Baja California blocks. In latest Cretaceous and earliest Tertiary time, the southerly elements of the margin moved north, not as a whole, but differentially, the Pacific Northwest, or Baja British Columbia as we may now call it, being carried outboard of the Klamath Mountains and southern Sierra Nevada to achieve its present position by Eocene time, and Baja California moving northwards to abut against the southern Sierra Nevada. During the mid-Tertiary, western and northern elements of Baja B.C. were displaced northwards forming the present Wrangel block (block 1 A of figure 6).

Umhoefer (1987) and Umhoefer et al. (1989 b) have developed a model that reconciles the above evidence with the motions of oceanic plates to the west (figure 6). Their model is based on the suggestion by Beck et al. (1981) that terranes were moved northwards by the short-lived Kula Plate (Atwater 1970). They assume that between 85 and 66 Ma that Baja B.C. was

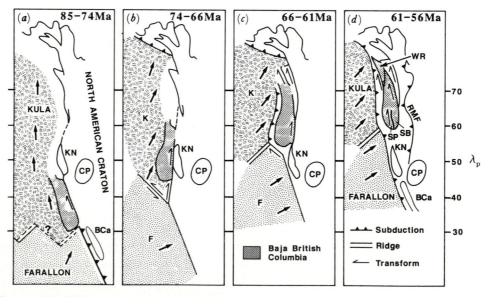

FIGURE 6. Plate model of Umhoefer (1987) and Umhoefer et al. (1989b) for evolution of the Cordillera in the interval 85–56 Ma. KN is the Klamath–Sierra Nevada; CP, Colorado Plateau; BCa, Baja California; WR, Wrangel Block; SP and SB, Spuzzum and Spences Bridge localities separated by transcurrent faults (Irving & Thorkelson 1990); RMF, Rocky Mountain Fault.

44

coupled to the Kula Plate. Plate reorganization at 66 Ma resulted in dextral oblique convergence between the Kula and North America plates west of Baja B.C. In the interval 66–55 Ma, Baja B.C. became detached from the Kula Plate continued to move northward relative to cratonic North America along inboard dextral faults. It was finally coupled to North America by 55 Ma.

6. Permian, Triassic, Jurassic

Determinations from bedded, palaeontologically well-dated, early Permian, late Triassic and early Jurassic rocks from Wrangellia and Stikinia are listed in table 2. In Triassic and Jurassic rocks, reversals occur, and their frequency is such that polarity zones cannot be related to the global timescale. Hence, from a palaeomagnetic datum it is not possible to determine whether the rock-unit was north or south of the palaeoequator. However, there are several reasons for believing that the palaeolatitudes, like those for Cretaceous rocks, are all northern. Firstly, Jurassic ammonite forms of the Canadian cordillera are boreal in character (Tipper 1981, 1984; Taylor *et al.* 1984; Cameron & Tipper 1985). Secondly, the Lower Permian Asitka Group (AS of table 2) has predominantly negative inclinations, the polarity expected for Northern Hemisphere localities in a geomagnetic field of reversed polarity, such as existed in the early Permian (Irving & Monger 1987). The third argument makes use of the fact that, with the exception of data from the Nicolai volcanics which have probably been moved northward during Tertiary coastwise motion (see below), the net displacements for the northern option are in excellent agreement with one another, differing by no more than 5°, which is not significant (figure 7). For the southern option they differ by as much as 30°. The Triassic and Jurassic rock-units from both Wrangellia and Stikinia are in stratigraphic sequence one above the other. That rock-units in stratigraphic continuity should have been separated by such large distances is most unlikely. Hence, the northern is the favoured option.

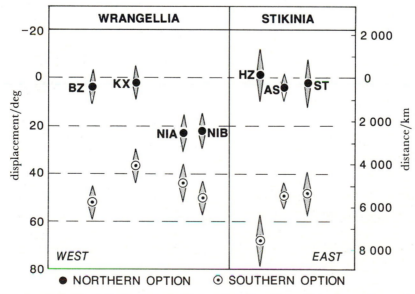

FIGURE 7. Relative palaeolatitudinal displacements of Lower Permian, Upper Triassic and Lower Jurassic localities from Wrangellia and Stikinia. Localities labelled as in table 2. Recent data from Lower Jurassic of central Stikinia confirms the displacements given here (Vandall *et al.* 1989).

Excluding data from the Nicolai, the net displacements do not differ significantly from zero. Apparently Stikinia and Wrangellia were close together in the late Triassic and early Jurassic, and were near their present latitudes relative to ancestral North America. However, palaeolongitude is indeterminate and positions westward in the Pacific are permissible.

The results of figure 7, although they agree in broad terms with other recent studies (Gordon *et al.* 1984; May & Butler 1986; Irving & Yole 1987; Irving & Wynne 1990; Beck 1990), differ from them in detail, as these studies do from one another. This arises because of the present unsatisfactory state of knowledge of the details Triassic and Jurassic APW path. Displacement estimates have not stabilized because data-selection and analytical procedures, adopted by different authors to establish cratonic reference data, differ substantially. Until choices are limited by more and better cratonic data, this somewhat unsatisfactory situation will persist. Nevertheless there seems little doubt that displacements are small, just how small remains to be seen. Changes are most likely to affect the displacement scale of figure 7 *b*, whereas the relative positions of points probably will not be grossly affected. In other words, the good agreement now found amongst palaeolatitude estimates for Lower Permian, Upper Triassic and Lower Jurassic rocks from Stikinia and the Vancouver Island segment of Wrangellia, is less vulnerable to future revisions than the net displacement values themselves.

We now must ask, how this good agreement was preserved while the later apparent displacements of figure 4 were occurring. Because of the uncertainty in the estimates of the early Mesozoic cratonic reference field, it is perhaps premature to attempt to answer, but one solution is to assume that the difference between the apparent displacements of the Coast and Intermontane belts was caused by variable tilting of plutons in the former (hypothesis (*c*) above).

Consider now the displacement of the Nicolai volcanics of Alaska which is about 20° (figure 7). They are essentially coeval with the Karmutsen Formation of Vancouver Island which yields zero net displacement. Palaeolatitudinally the two, however, are in close accord (figure 8). The corresponding rotations are very large (almost 180° in the case of Vancouver Island) and variable. Evidently, in the late Triassic, Wrangellia was a compact terrane that was later fragmented and the Alaskan element moved north along the western margin of North America.

The rotations required are large, variable and well documented. The geometrical relationship of locality and APW path is such that estimates of rotations are essentially independent of selections used to derive cratonic reference palaeopoles. It is important to note that although the net rotations are well established, the magnitude and, to a degree, the sense of post-depositional pre-mid-Cretaceous rotations estimated are much less certain, because they depend on the amount of the post-mid-Cretaceous dextral rotation assumed for the localities in question. It has not yet proved possible, and may never be possible because of the structural complexity of Western Cordillera, to measure both the early sinistral and later dextral rotation in the same uninterrupted sequence.

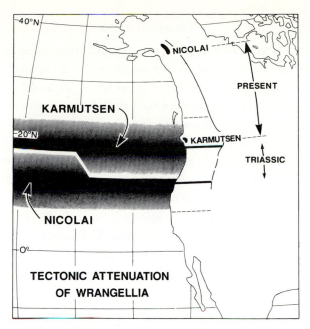

FIGURE 8. Palaeolatitudes from Upper Triassic rocks of Wrangellia compared. Details in table 2. Note that the error zones are shaded densely near the mean and with the density diminishing outwards to the limit ($P = 0.05$).

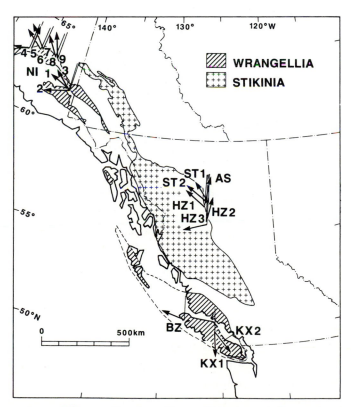

FIGURE 9. Apparent rotations of Lower Permian, Upper Triassic and Lower Jurassic localities, Northern Hemisphere option. Localities labelled as in table 2. Large anticlockwise rotations have recently been reported from Lower Jurassic rocks of central Stikinia (Vandall *et al.* 1989).

47

7. Conclusions

Consider again the three processes invoked to explain data from Cretaceous rocks of the Pacific Northwest: (a) general 30° tilt to the southwest, (b) displacement from the south increasing from east to west, greater than 2000 km in the west, and (c) lesser (1000–2000 km) displacement from the south, and within the Coast Belt, lesser tilts to the south and southeast. The displacement estimates obtained from a wide variety of bedded rocks of Permian through Jurassic age are in full agreement with hypothesis (a), and together they indicate that the Cordillera of the Pacific Northwest was formed initially more or less where it is now. Data from bedded volcanics of the Intermontane Belt are inconsistent with hypothesis (a), but provide support for hypotheses (b) and (c). Strong support for the process of coast-wise displacement in general is provided by Cretaceous palaeomagnetic data from Baja California (figure 5) and, of course, by the present motions of western and Baja California relative to North America; although these have no direct bearing on the tilt-displacement problem in the Pacific Northwest, they do show that coast-wise displacements of the margins of North America are not isolated phenomenon.

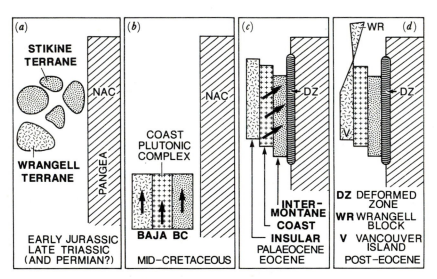

FIGURE 10. Speculative scheme for evolution of the allochthonous terranes of Western Canada. NAC is the North American craton. DZ is the deformed zone consisting of Omineca and Foreland belts with the possibility that there may have been some involvement of the inner part of the Intermontane Belt. The heavy arrows in the centre two panels represent diagrammatically the directions of palaeonorth that were systemmatically rotated during attenuation and differential northward movement. The northward slivering of the Wrangel block in the right-hand panel is meant to represent the final attenuation of the Insular Belt in the Tertiary (see figure 9).

Hypothesis (a) is summarized in figure 10. The various terranes that now constitute the Insular, Coast and Intermontane belts were originally situated close together, and not far from where they are at present relative to ancestral North America. This choice is arbitrary to a degree, because the indeterminacy of longitude allows positioning farther west. Uncertainties in the cratonic reference field are such that they could have been situated several hundred kilometres to the north or south. Presumably Wrangellia and Stikinia in the late Triassic and early Jurassic were, in large part, a succession of island-arcs, seamounts, etc. They moved southward together, became amalgamated and attached temporarily to North America in

Jurassic through early Cretaceous time. Large rotations, predominantly anticlockwise, occurred. In mid-Cretaceous time they were intruded by the Coast Plutonic Complex and Baja B.C. was formed. In the latest Cretaceous and earliest Tertiary (90–50 Ma), Baja B.C. moved northward along the margin of North America to its present position and clockwise rotations commonly occurred. Finally the Wrangel block was displaced as a sliver 20° northward.

The displacement hypothesis of figure 10 depends, perhaps to a disproportionate degree, on the systematic nature of the westerly increase in apparent displacements estimated from mid-Cretaceous (120–102 Ma) plutons, and the displacement estimated from the stratified mid-Cretaceous (104 Ma) Spences Bridge Group (figure 4). Clearly very much more data are required from this time interval before the thesis of figure 10 can be considered as other than tentative. The displacement estimate from the latest Cretaceous (70 Ma) Carmacks Group (Marquis & Globerman 1988) also is consistent with this thesis, but as already noted, is based on an as yet uncertain reference field. However, the Carmacks datum is not critical, because the northward movement of Baja B.C. may have been partly or even largely accomplished by this time.

Appendix

Data from Mesozoic rocks, other than those listed in table 2, are available from the Pacific Northwest but have not been included in our analysis for reasons given below.

The Cretaceous plutons of Howe Sound (Symons 1973a) have been shown to contain substantial Tertiary overprints (Irving et al. 1985; E. Irving & C. J. Yorath, unpublished work). Cretaceous rocks of the Methow–Pasayten trough have syn-deformational magnetizations that cannot be related confidently to palaeohorizontal (Granirer et al. 1986). Late Cretaceous sedimentary rocks of McColl Ridge (Panuska 1985) of the Wrangel Block, although providing a positive tilt test from two localities, are based on only a few sedimentary rock samples which may have been affected by inclination error (Coe et al. 1985). Data from the Topley intrusions (Symons 1973b, 1983a) of Stikinia are affected by early Tertiary overprinting (Monger & Irving 1980) in our opinion. Despite their shortcomings, these data, if accepted at their face-value, would all yield southerly displacements, broadly consistent with interpretation given above.

Data from the Albian Crowsnest Formation are not considered because the unit probably represents a single eruption insufficient to average the palaeosecular variation (Irving et al. 1986).

The Triassic Hound Island volcanics (Hillhouse & Grommé 1980) from the Alexander terrane have been restudied by Haeuseller et al. (1989), who report that the earlier determination was affected by Cretaceous overprinting, and that detailed demagnetization yields magnetizations with inclinations concordant with the contemporaneous Nicolai volcanics.

Upper Triassic volcanics and Lower Jurassic intrusions of south central British Columbia (Quesnellia) have yielded magnetizations directed towards the northeast (GBT, GBA, CMT, CMA, NA of figure 11) with positive inclination. Similar palaeodirections, known to be overprints, have been observed in nearby Norian to Hettangian volcaniclastics (QL). On Vancouver Island, comparable palaeodirections have been observed from the Jurassic West Coast Complex and Island Intrusions (WI), and as overprints from the Karmutsen Formation

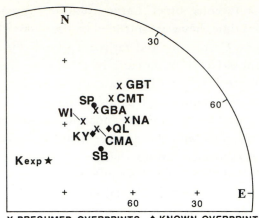

X PRESUMED OVERPRINTS ♦ KNOWN OVERPRINTS
● MID–CRETACEOUS MAGNETIZATIONS
★ EXPECTED FROM CRATON

FIGURE 11. Examples of magnetizations in Triassic and Jurassic rock-units that are interpreted as entirely or in part of mid-Cretaceous age. Directions plotted on stereogram are recalculated to the town of Kamloops (50.6° N, 120.4° W) for comparison. All directions with respect to present horizontal except SB. WI, West Coast Complex and Island intrusions (Symons 1984a); CMT (Symons & Litalien 1984), CMA (Symons 1973a), Copper Mountain intrusion derived from thermal and alternating field demagnetization studies; GBA (Symons 1971a), GBT (Symons 1983b), Guichon Batholith derived from alternating field and thermal studies; KY, Karmutsen Formation (Yole & Irving 1980); NA, Nicola volcanics (Symons 1984b); QL, Quesnel lake volcaniclastics overprint (Rees *et al.* 1985); SP and SB, Spuzzum pluton and Spences Bridge Group from table 2. Expected mid-Cretaceous palaeodirection (K_{exp}) calculated from data of table 1.

(KY). All are broadly similar to known Cretaceous palaeodirections (SB, SP). The sampling area in south-central British Columbia has been the focus of extensive mid-Cretaceous volcanism and intrusion. Although statistically significant differences occur among the palaeodirections of figure 11, the overall agreement is such that we feel justified in concluding that these rock-units have been partly or completely remagnetized in mid-Cretaceous time, and hence, cannot be used confidently for earlier tectonic reconstructions. Note, however, that all are markedly different in direction from the expected Cretaceous palaeofield.

REFERENCES

Alvarez, W., Kent, D. V., Premoli, S. I., Schweichert, R. A. & Larson, R. L. 1980 *Bull. Geol. Soc. Am.* **91**, 476–484.
Armstrong, R. L., Monger, J. W. H. & Irving, E. 1985 *Can. J. Earth Sci.* **22**, 1217–1222.
Atwater, T. 1970 *Bull. Geol. Soc. Am.* **81**, 3513–3536.
Bardoux, M. & Irving, E. 1989 *Can. J. Earth Sci.* **22**, 829–844.
Beck, M. E. Jr 1975 *Earth planet. Sci. Lett.* **26**, 263–268.
Beck, M. E. Jr 1976 *Am. J. Sci.* **276**, 694–712.
Beck, M. E. Jr 1980 *J. geophys. Res.* **85**, 7115–7131.
Beck, M. E. Jr 1990 In *Geophysical framework of the Continental United States* (ed. L. C. Pakiser & W. D. Mooney). *Geol. Soc. Am. Mem.* **172**. (In the press.)
Beck, M. E., Burmester, R. F. & Schoonover, R. 1981 *Earth planet. Sci. Lett.* **56**, 336–342.
Beck, M. E. & Noson, L. 1972 *Nature, Lond.* **235**, 11–13.
Bogue, S. W., Cowan, D. S. & Garver, J. I. 1989 *J. geophys. Res.* **94**, 10415–10427.
Butler, R. F., Harms, T. A. & Gabrielse, H. 1988 *Can. J. Earth Sci.* **25**, 1316–1322.
Butler, R. F., Gehrels, G. E., McClelland, W. C., May, S. R. & Klepacki, D. 1989 *Geology* **17**, 691–694.
Cameron, B. E. B. & Tipper, H. W. 1985 *Bull. Geol. Surv. Can.* **365**, 1–49.
Champion, D. E., Howell, D. G. & Grommé, C. S. 1984 *J. geophys. Res.* **89**, 7736–7752.
Champion, D. E., Howell, D. G. & Marshall, M. 1986 *J. geophys. Res.* **91**, 11557–11570.

Coe, R. S., Globerman, B. R., Plumley, P. W. & Thrupp, G. A. 1985 In *Tectonostratigraphic terranes of the Circum-Pacific region* (ed. D. G. Howell), Circum-Pacific Council for Energy and Mineral Resources, pp. 85–108. Houston, Texas.

Courtillot, V., Feinberg, M., Ragaru, J. P., Kerguelen, R., McWilliams, M. & Cox, A. 1985 *Geology* **13**, 107–110.

Cox, A. V. 1957 *Nature, Lond.* **179**, 685–686.

Cox, A. V. 1980 *Geol. Ass. Can., Spec. Pap.* **20**, 305–321.

de Boer, J. & Snider, F. G. 1979 *Bull. Geol. Soc. Am.* **90**, 185–198.

Dooley, R. E. & Smith, W. A. 1982 *Tectonophysics* **90**, 283–307.

Filmer, P. E. & Kirschvink, J. L. 1989 *J. geophys. Res.* **94**, 7332–7342.

Fox, K. F. & Beck, M. E. 1985 *Tectonics* **4**, 323–341.

Frei, L. S., Magill, J. R. & Cox, A. 1984 *Tectonics* **3**, 157–177.

Frei, L. S. 1986 *Bull. Geol. Soc. Am.* **97**, 840–849.

Fry, J. G., Bottjer, D. J. & Lund, S. P. 1985 *Geology* **13**, 648–651.

Gabrielse, H. 1985 *Bull. Geol. Soc. Am.* **96**, 1–14.

Geissman, J. W., Collins, J. T., Oldow, J. S. & Humphries, S. R. 1984 *Tectonics* **3**, 179–200.

Globerman, B. R. & Irving, E. 1988 *J. geophys. Res.* **93**, 11721–11733.

Gordon, R. G., Cox, A. & O'Hare, S. 1984 *Tectonics* **3**, 499–537.

Granirer, J. L., Burmester, R. F. & Beck, M. E. Jr 1986 *Geophys. Res. Lett.* **13**, 733–736.

Grommé, S. 1984 *Pacific Section S.E.P.M.* **43**, 113–119.

Grommé, C. S. & Merrill, R. T. 1965 *J. geophys. Res.* **70**, 3407–4320.

Grubbs, K. L. & van der Voo, R. 1976 *Tectonophysics* **33**, 321–336.

Hagstrum, J. T., McWilliams, M., Howell, D. G. & Grommé, S. 1985 *Bull. Geol. Soc. Am.* **96**, 1077–1090.

Harland, W. B., Cox, A. V., Hewellyn, P. G., Pickton, C. A. G., Smith, A. G. & Walters, R. 1982 *A geological time scale*. Cambridge University Press. (131 pages.)

Hillhouse, J. W. 1977 *Can. J. Earth Sci.* **14**, 2578–2592.

Hillhouse, J. W. & Grommé, C. S. 1980 *J. geophys. Res.* **85**, 2594–2602.

Hillhouse, J. W. & Grommé, C. S. 1982 *Geology* **10**, 552–556.

Hillhouse, J. W. & Grommé, C. S. 1984 *J. geophys. Res.* **89**, 4461–4477.

Hodych, J. & Hayatsu, A. 1988 *Can. J. Earth Sci.* **25**, 1972–1989.

Haeussller, P. J., Coe, R. S. & Onstott, T. C. 1986 *Eos, Wash.* **70**, 1068.

Irving, E. 1985 *Nature, Lond.* **314**, 673–674.

Irving, E. & Archibald, D. A. 1990 *J. geophys. Res.* (In the press.)

Irving, E. & Brandon, M. T. 1990 *Can. J. Earth Sci.* (In the press.)

Irving, E. & Irving, G. A. 1982 *Geophys. Surv.* **5**, 141–188.

Irving, E. & Monger, J. W. H. 1987 *Can. J. Earth Sci.* **24**, 1490–1494.

Irving, E., Monger, J. W. H. & Yole, R. W. 1980 *Geol. Ass. Can. Spec. Pap.* **20**, 441–456.

Irving, E. & Thorkelson, D. K. 1990 *J. geophys. Res.* (Submitted.)

Irving, E., Woodsworth, G. H., Wynne, P. J. & Morrison, A. 1985 *Can. J. Earth Sci.* **22**, 584–598.

Irving, E. & Wynne, P. J. 1990 In *The Cordilleran Orogen in Canada* (ed. H. Gabrielse & C. J. Yorath). DNAG contribution. (In the press.)

Irving, E., Wynne, P. J., Evans, M. E. & Gough, W. 1986 *Can. J. Earth Sci.* **23**, 591–598.

Irving, E. & Yole, R. W. 1972 *Pub. Earth Phys. Branch* **42**, 87–95.

Irving, E. & Yole, R. W. 1987 *Geophys. Jl R. astron. Soc.* **91**, 1025–1048.

Jacobson, D., Beck, M. E., Diehl, J. K. & Hearn, B. C. 1980 *Geophys. Res. Lett.* **7**, 549–552.

Kanter, L. R. & McWilliams, M. O. 1982 *J. geophys. Res.* **87**, 3819–3830.

King, R. F. 1955 *Mon. Not. R. astr. Soc. geophys. Suppl.* **7**, 115–134.

Mankinen, E. A. & Irwin, W. P. 1982 *Geology* **10**, 82–87.

Marquis, G. & Globerman, B. R. 1988 *Can. J. Earth Sci.* **25**, 2005–2016.

May, S. R. & Butler, R. F. 1986 *J. geophys. Res.* **91**, 11519–11544.

McWilliams, M. O. & Howell, D. G. 1982 *Nature, Lond.* **297**, 215–217.

Monger, J. W. H. 1990 In *The Cordillera in Canada* (ed. H. Gabrielse & C. J. Yorath). DNAG. (In the press.)

Monger, J. W. H. & Irving, E. 1980 *Nature, Lond.* **285**, 289–294.

Monger, J. W. H., Price, R. A. & Tempelman-Kluit, D. J. 1982 *Geology* **10**, 70–75.

Packer, D. R. & Stone, D. B. 1974 *Can. J. Earth Sci.* **11**, 976–997.

Palmer, A. R. 1983 *Geology* **11**, 503–504.

Panuska, B. C. 1985 *Geology* **13**, 80–883.

Price, R. A. & Carmichael, D. M. 1986 *Geology* **14**, 468–471.

Rees, C. J., Irving, E. & Brown, R. L. 1985 *Geophys. Res. Lett.* **12**, 498–501.

Russell, B. J., Beck, M. E. Jr, Burmester, R. F. & Speed, R. C. 1982 *Geology* **10**, 423–428.

Schwarz, E. J., Muller, J. E. & Clark, K. R. 1980 *Can. J. Earth Sci.* **17**, 389–399.

Smith, T. E. & Noltimier, H. C. 1979 *Am. J. Sci.* **279**, 778–807.

Symons, D. T. A. 1971*a* *Can. J. Earth Sci.* **8**, 1388–1396.

Symons, D. T. A. 1971*b* *Geol. Surv. Can. Pap.* **71–24**, 9–24.

Symons, D. T. A. 1973a *Nature, Lond.* **241**, 59–61.
Symons, D. T. A. 1973b *Can. J. Earth Sci.* **10**, 1099–1108.
Symons, D. T. A. 1977 *Can. J. Earth Sci.* **14**, 2127–2139.
Symons, D. T. A. 1983a *Geophys. Res. Lett.* **10**, 1065–1068.
Symons, D. T. A. 1983b *Can. J. Earth Sci.* **20**, 1340–1344.
Symons, D. T. A. 1984a *J. Geodyn.* **2**, 229–244.
Symons, D. T. A. 1984b *J. Geodyn.* **2**, 211–228.
Symons, D. T. A. & Litalien, C. R. 1984 *Geophys. Res. Lett.* **11**, 685–688.
Symons, D. T. A. & Wellings, M. R. 1989 *Can. J. Earth Sci.* **26**, 821–828.
Tarduno, J. A., McWilliams, M., Debiche, M. G., Sliter, W. V. & Blake, M. C. 1985 *Nature, Lond.* **317**, 345–347.
Tarduno, J. A., McWilliams, M., Sliter, W. V., Cook, H. E., Blake, M. C. Jr & Premoli-Silva, I. 1986 *Science, Wash.* **231**, 1425–1428.
Taylor, D. G., Callonon, J. H., Hall, R., Smith, P. L., Tipper, H. W. & Westermann, G. E. G. 1984 *Geol. Ass. Can. Spec. Pap.* **27**, 121–141.
Teissere, R. F. & Beck, M. E. Jr 1973 *Earth planet. Sci. Lett.* **18**, 296–300.
Tipper, H. W. 1981 *Can. J. Earth Sci.* **18**, 1788–1792.
Tipper, H. W. 1984 *Geol. Ass. Can. Spec. Pap.* **27**, 113–120.
Umhoefer, P. J. 1987 *Tectonics* **6**, 377–394.
Umhoefer, P. J., Granier, J. I., Schiarizza, P. & Glover, J. K. 1989a *Geol. Soc. Am. Cordilleran Meeting* (abst.), p. 92.
Umhoefer, P. J., Dragovich, J., Cary, J. & Engebretson, D. C. 1989b *Geophys. Monog. Am. Geophys. Union* **50** (ed. J. W. Hillhouse), pp. 101–111.
Vandall, T. & Palmer, H. C. 1988 *Eos, Wash.* **69**, 1165.
Vandall, T., Palmer, H. C. & Woodsworth, G. J. 1989 *Abst. Geophys. Union 16th A. Mtg* (Abst.), no. 77.
Witte, W. K. & Kent, D. V. 1989 *Bull. Geol. Soc. Am.* **101**, 1118–1126.
Wu, F. & Van der Voo, R. 1988 *Tectonophysics* **156**, 51–58.
Yole, R. W. & Irving, E. 1980 *Can. J. Earth Sci.* **17**, 1210–1228.

Discussion

P. F. HOFFMAN (*Geological Survey of Canada, Ottawa*). Granitic rocks deform by dislocation creep at temperatures well below the Curie temperature. How, then, does one determine the local palaeohorizontal for palaeomagnetic purposes?

E. IRVING, F.R.S. Within the regional groups of figure 5, the remanence directions in plutons are in good agreement with one another and with those of bedded rocks where available. We argue that this consistency observed between rocks of very different origins rules out local processes, such as dislocation creep or tilting, as major causes of the aberrances observed in figure 4. Under favourable circumstances, local palaeohorizontal for individual plutons can be obtained by estimating the tilt of bathozones in their vicinity, as noted in the text.

A. TRENCH (*Department of Earth Sciences, University of Oxford, U.K.*). Professor Irving demonstrated that palaeomagnetic anomalies in both inclination and declination exist between the terranes of the Western Canadian Cordillera and coeval strata on the North American craton. Clockwise declination anomalies are linked to dextral movements in the plate boundary zone whereas anti-clockwise anomalies are found to accompany sinistral movements. To what extent do intra-terrane declination variations exist within the Cordillera? If present, do these record temporal evolution of rotation or do they result from local tectonic rotations within a given terrane?

A. H. F. ROBERTSON (*Grant Institute of Geology, Edinburgh, U.K.*). Is there any evidence of small-scale block rotations within the transported terranes, especially near the bounding strike–slip faults, or do they behave as relatively coherent units?

E. IRVING, F.R.S. Apparent dextral rotations of 40–70° observed in Cretaceous rocks are always associated with apparent northward displacement (figure 5). The hypothesized northward movement of Baja B.C. attached to the Kula Plate would produce an *en bloc* dextral rotation relative to North America of about 20°. Additional rotations could have been caused by local block rotations (Umhoefer 1987), but the structures defining these have not been identified.

Variable but predominantly sinistral net rotations occur in Upper Triassic and Lower Jurassic rocks. Because these have all been observed from bedded rocks, they are the best established and in some ways the most intriguing palaeomagnetic results from the Cordillera. Their common occurrence indicates that they are present throughout the terranes involved, and are not just a marginal phenomenon (figure 9). For example, in Stikinia the bunching of observations near the eastern margin appears to be a sampling artifact, because recently Vandall *et al.* (1989) have reported, in abstract, variable net sinistral rotations of 20 to 110° from Lower Jurassic rocks of central Stikinia. The timing of rotations observed from Triassic and Jurassic rocks, is poorly constrained but they are so widespread that I am inclined to believe that they occurred soon after deposition as a consequence of rifting in a transcurrent or transpressive environment.

On terrane analysis

By W. B. Hamilton

U.S. Geological Survey, Denver, Colorado 80225, *U.S.A.*

Many recent papers on displaced crustal masses are organized around 'terranes' given only geographic names, and are bewildering for other than local experts; self-explanatory descriptive or genetic terms should be incorporated in most designations. Plate-tectonic interpretations of aggregated terranes that incorporate awareness of how modern arc systems vary and evolve contain implicit and testable predictions, but too many interpretations are based instead on invalid assumptions. As actualistic models are applied to analyses of orogenic belts, much more mobilistic interpretations than those generally now visualized will probably emerge. An example is made of the Carpathian region.

1. Introduction

Most orogenic belts contain as important components the sweepings from subducted oceanic plates and the variably aggregated and deformed light crustal masses brought together as plates converged by subduction of intervening oceanic lithosphere. Many belts further display disruption by strike–slip faulting. The characterization of rock assemblages juxtaposed by such processes, and the evaluation of palaeomagnetic, palaeontologic, and palaeoenvironmental evidence for motion histories, are critical steps in deciphering the evolution of orogenic belts. Much additional understanding of orogenic belts can come from the seeking of plate-tectonic relationships between components.

2. Terrane nomenclature

All orogenic belts contain tracts that have been transported long distances relative to adjacent tracts from which they are now separated by sutures or major faults. Coney *et al.* (1980) proposed that such tracts be termed 'terranes', and the term has become popular (see, for example, Howell *et al.* 1985; Howell 1989) for known ('allochthonous', 'displaced', or 'exotic' terranes) or possible ('suspect' terranes) transported masses. The term often is given as 'tectonostratigraphic terrane'. Different geologists have formally named terranes as diverse as out-of-place continental masses or island arcs, or lithologic units therein; entire accretionary wedges, or contrasted melange assemblages, or isolated slices of coherent material in wedges; displaced bits in strike–slip régimes; adjacent thrust sheets in foreland thrust belts; components within crystalline complexes that might be bounded by faults; and concordant metavolcanic and metasedimentary packages in the walls of a pluton.

The best place to see how the term terrane is used is in the volume edited by Howell (1985), wherein most authors apply it to lithologic or tectonic units which might have been juxtaposed by plate motions. One group of authors designates terranes with geographic names and organizes descriptions around these. Ji & Coney (1985) split China into 33 'fault bounded' terranes and schematize the speculative Phanerozoic history of each with a column of lithology and age. Most of these terranes are inferred to be continental fragments; ages of collisions are asserted for all terranes, although few sutures are known, few bounding structures are characterized, and some boundaries are merely basin onlaps. More soundly based, but similarly organized around geographically named terranes, are five papers (Blake *et al.* 1985 *a, b*; Hill 1985; Irwin 1985; Mortimer 1985) in the volume dealing with the Klamath Mountains region of California and Oregon. These papers collectively name 28 terranes and subterranes, plus numerous additional 'belts', all without qualifying nouns or adjectives regarding age, lithology, structure, or tectonic setting. Descriptions are keyed to these memory-defeating abstract names. Despite the obvious relevance for plate-tectonic analysis of the melanges, island arcs, and ophiolites hidden in the nomenclature of these Klamath classifiers, none of the five papers attempts plate-tectonic synthesis. None tries to pair accretionary wedges to magmatic arcs, or to define polarities of subduction, or to evaluate whether sutured arcs were accreted separately to the continent or were aggregated first into composite island arcs. Only a few units and structures are even assigned vague plate-tectonic contexts with single phrases or sentences. There are mentions of palaeomagnetic and palaeontologic evidence for plate motions but no analyses of how these were accommodated.

By contrast, one paper (Harper *et al.* 1985) in the Howell volume presents a plate-tectonic synthesis of the Klamath region, and instead of terrane terminology, it uses an easily understood, descriptive terminology, in part genetic, for the components of the complexly sutured aggregate. Other authors in the volume who present plate-tectonic syntheses for other regions and who either eschew terrane terminology in favour of descriptive nomenclature or add descriptive modifiers to terrane terms include Box (1985), Ernst *et al.* (1985), Hawkins *et al.* (1985), Kimbrough (1985), and McCabe *et al.* (1985).

These and other recent papers display a tendency for investigators of displaced crustal masses to place themselves in one of two contrasted groups. Most of those who favour geographic terrane nomenclature espouse objective description and classification and largely neglect genetic relationships between assemblages; predictions are not made and hence few tests are applied. Most members of the opposed descriptive-nomenclature group use models of genetic relationships and advance plate-tectonic interpretations through self-explanatory terminology. Predictions implicit in the model-driven classifications can be tested, which permits rejection of inappropriate models.

It appears that the proportion of researchers who understand the empirical behaviour and relationships of components of modern plate-interaction systems is much higher in the descriptive-terminology group than in the geographic-terminology one. Such understanding is not easily obtained, for not only does no textbook contain it, but most of the casual treatments of plate-boundary systems that do appear in textbooks incorporate false assumptions. Even the new book by Howell (1989) lacks understanding both of subduction and of triple junctions. My harsh inference is that members of the geographic-terminology group do not present the critical evidence needed for evaluating plate interactions because they generally do not comprehend how plates interact.

Howell (1989) argued that doubts as to the reality of additional terranes should be resolved by naming all possible ones, reserving for later investigators the task of determining whether or not adjacent terranes have been juxtaposed across major structures. Application of this philosophy will not only clutter the literature but, worse, will inevitably lead to circular rationales justifying the initial discriminations. Indeed, much foolish speculation already has come from the practice, in northern Alaska and southern California, for example.

The following section notes some of the features of modern plate systems that should be taken into account when progressing from classification to comprehension of ancient orogenic belts. Few geologists working with orogenic belts comprehend actualistic plate tectonics.

3. ACTUALISTIC PLATE TECTONICS

Plate tectonics has given us the framework within which to perceive relationships between tectonic and magmatic features and movement histories. Conversely, plate-tectonic interpretations carry implicit predictions that can be tested against other data. Much of the crustal material in orogenic belts has been tectonically accreted by processes related to subduction. Such material was conveyor-belted atop plates of oceanic lithosphere toward subduction zones, at which some materials disappeared beneath overriding plates whereas others, variably dismembered, were scraped off against them. Such accretion can take place against an island arc or an Andean-type continental margin, or between colliding arcs. As overriding continental plates are themselves moving, with their bounding subduction systems, and as island arcs migrate, reverse polarity, and collide with each other and with continents or lesser crustal masses, histories and characteristics of accreted orogenic complexes commonly vary greatly along strike. Accreted materials can have moved 10 000 km relative to the continent against which they are emplaced; or they may be rifted away a mere 100 km, then collapsed back by subduction. Sedimentation in trenches is mostly by turbidites flowing along the trenches, rather than by sediments moving directly into trenches from adjacent arc sectors, and terrigenous sediments can have sluiced 3000 km from their sources before being imbricated into an accretionary wedge.

Seven very large lithospheric plates, and numerous mid- and small-sized ones (the concept of coherent plates breaks down at the small-scale end), are now all moving relative to all others. Corollaries are that all plate boundaries are also moving, and that most change greatly in length and shape as they move. Parts of many plates undergo severe internal deformation, and boundaries often jump or migrate to new positions. Relative velocities between adjacent plates presently range up to about 15 cm a^{-1}.

Arc systems develop where oceanic plates sink beneath overriding plates that can be continental, transitional, or oceanic. Many individual arcs are continuous across the diverse crustal types. Arcs commonly are inaugurated by subduction reversals consequent on collisions between other arcs and light crustal masses, and collision histories vary greatly along trend. Oceanic arcs migrate and lengthen with time, and one sector of a continuous arc can have been inaugurated tens of millions of years later than another sector, or after another sector has collided and stopped. Petrologic and crustal features evolve as activity continues in a given sector. I discussed and referenced such features in recent papers (Hamilton 1988, 1989) and documented many of them in a monograph (Hamilton 1979).

'Absolute' velocities of present large plates – their relative velocities in an approximate zero-sum frame – correlate positively with the lengths of ridges and of trenches along their perimeters, and negatively with the proportion of continental lithosphere within them (Carlson 1981). Plates are propelled primarily by gravitational forces, and on average pull by the descending slab is about 2.5 times as important in moving plates as is slide of plates away from ridges, whereas thick continental lithosphere retards motion by drag (Carlson 1981). Subduction-pull increases with density and thickness, and hence with age, of subducting lithosphere, and forward velocities of subducting plates, roll-back velocities of trench hinges, and extension or drag of overriding plates vary correspondingly (Hilde & Uyeda 1983; Jarrard 1986; Jurdy & Stefanick 1988). The 80 or 100 km of relief of the base of an oceanic plate, between lithosphere and less-dense asthenosphere, is much more important in producing ridge-slide than is the 3 or 4 km of bathymetric relief of the top of the plate. Major plate motions are controlled primarily by large lateral variations in lithosphere density and thickness that result primarily from cooling (Carlson 1981; Hager & O'Connell 1981).

Much published tectonic speculation and geophysical modelling incorporates the false assumptions that subducting plates roll over hinges, and slide down slots, that are fixed in the mantle, and that overriding plates commonly are shortened compressively. Rather, hinges commonly retreat – roll back – into incoming oceanic plates as overriding plates advance, even though at least most subducting plates are also advancing in 'absolute' motion. Subducting slabs sink more steeply than the inclinations of Benioff seismic zones, which mark positions, not trajectories, of slabs. Perhaps the most obvious evidence for roll-back is that the Pacific Ocean is becoming smaller with time, but much other evidence has been presented by many authors. The typical régime in an overriding plate above a sinking slab is one of extension, not shortening (although of course great shear and shortening affect the accretionary wedge in front of the overriding plate). A corollary, often overlooked in palinspastic analyses, is that subduction can occur beneath only one side at a time of an internally rigid plate.

As island arcs migrate oceanward over subducting plates, marginal basins of new oceanic lithosphere open behind the arcs (Taylor & Karner 1983). Some migration is accomplished by the splitting of a magmatic arc and the migration of its forward half away from the rear half, and some by irregular sea-floor spreading behind the entire arc. The magmatic welt can move forward with the advancing part of the overriding plate, can be abandoned as a remnant arc on the rear plate, or can be split between them. Oceanic island arcs do not bound rigid plates of old lithosphere, but instead mark the fronts of plates of young lithosphere that are widening in the extensional régimes above sinking slabs.

Arcs increase in curvature as they migrate. A migrating arc becomes pinned where it encounters thick crust in the subducting plate that either is non-subductible or that forms a stiffening girder, and festoons and sharply curving arcs result where migration continues away from such obstructions.

An island arc is not a feature fixed on an overriding plate, but rather is a belt of arc-magmatic rocks formed above that part of a subducting slab whose top is 100 km or so deep. The belt migrates, with or without the part of the overriding plate on which it stands, to track that contour as the slab falls away. Although much oceanic back-arc-basin lithosphere forms by regular or irregular spreading behind an arc, much also may form by the rapid migration of a magmatic arc which forms a sheet of arc crust rather than forming a full island-arc welt. On-land ophiolites are sections of upper oceanic lithosphere that probably are dominated by these two types of back-arc basin materials rather than by the products of spreading mid-ocean ridges.

There appear to be two major processes of emplacement of ophiolites within orogenic belts. One is in the collision of an advancing arc with a continent or other island arc, the thin ophiolitic leading edge of the overriding plate ramping onto thick-crustal parts of subducting plates. Such thrusting is in the sense of subduction. (The hypothetical process of 'obduction', whereby a sheet of oceanic lithosphere is split from a subducting slab and shoved, in the opposite sense of thrusting, atop the thick crust of an overriding island-arc or continental plate, is invoked by many writers, but the process defies mechanical analysis and has yet to be proved to have operated anywhere. Confusion is introduced by writers who misuse the term 'obduction' to imply onramping in the sense of subduction, which does happen but is opposite to the original definition of the term.)

The second major process of ophiolite emplacement is a common byproduct of collision of an arc with a continent or another arc. A new subduction system of opposite dip breaks through behind the collided arc and leaves attached to it a strip of back-arc basin crust that becomes raised as accretionary-wedge materials are stuffed beneath it. Such an ophiolitic strip may remain at the leading edge of a plate as long as subduction continues beneath it, subject to tectonic erosion, but it will become part of a suture system when another non-subductible crustal mass collides with it. Oceanic arcs commonly are not inaugurated by the breaking of subduction through old oceanic crust, but rather break through near boundaries between thin and thick crust and migrate over the plates of thin crust (Hamilton 1979, 1988; Karig 1982).

4. Palaeotectonic analysis

The evaluation of ancient tectonic systems in terms of terranes and actualistic plate tectonics has barely begun. Shortcomings of terrane analysis were noted earlier, and on the other hand too many published analyses of orogenic belts, although purportedly plate-tectonic, are made without understanding of plate behaviour. The success of the plate-tectonic paradigm has produced a complacency that has cluttered the geologic literature with speculations based on invalid models and devoid of testable predictions.

The central European belt of complex late Mesozoic and Cenozoic terrane accretion has been intensively studied by generations of geoscientists yet still is understood only poorly. Interpretations continue to vary widely but have tended to evolve from geosynclinal toward plate-tectonic concepts. The general view is that the region records the slow opening and closing of small ocean basins and the accompanying collisions of crustal fragments, but all details are disputed. Recent analyses range from modest modifications of stabilist theory (Debelmas 1989; Foldvary 1988; Sandulescu 1988), wherein slow subduction of small oceanic basins is added to classical Alpine overthrusting explanations, to sophisticated interpretations that incorporate actualistic plate tectonics (Dewey et al. 1989; Gealey 1988).

I view Europe from the perspective of Indonesian and western Pacific analogues (Hamilton 1989) and regard even the most mobilistic of the modern interpretations of Alpine Europe as likely too conservative. If those analogues are indeed appropriate for evaluating the broad and complex Tethyan region of Mesozoic and Cenozoic rifting and plate convergence, then Alpine Europe may record the subduction of much more oceanic lithosphere than is commonly assumed. Too little attempt has yet been made to pair accretionary wedges to magmatic arcs, or to match palaeobiogeographic and sedimentological data to rifting and drifting histories, or to entertain complexities like those now known to operate in Pacific arcs.

Within central Europe, my own familiarity is greatest with the Carpathian system, where the distribution of late Mesozoic and Cenozoic accretionary-wedge materials, magmatic-arc igneous rocks, and other plate-interaction features lead me to a far more mobilistic view than is held by most regional experts. (My familiarity comes partly from the literature, and partly from extensive field-tripping in Slovakia, organized by Čestmír Tomek and led by regional experts, and from access to unpublished data and discussions with many geoscientists on a visit to Hungary coordinated by György Pogácsás.) My analysis shares much more with the early plate-tectonic papers by Bleahu *et al.* (1973) and Szadeczky-Kardoss (1973) than with most more recent syntheses of the regional geology (Burchfiel 1976; Csaszar *et al.* 1987; Foldvary 1988; Kazmer & Kovacs 1985; Royden & Baldi 1988; Sandulescu 1988; Unrug 1984).

As most observers agree, the Carpathian arc apparently is a response to the tectonic filling of what was in early Tertiary time a westward-opening oceanic embayment. The presently dominant explanations for the tectonic filling (Balla 1986; Royden & Baldi 1988; Royden & Burchfiel 1989) visualize the jamming ('escape tectonics') of a previously coherent continental plate, pushed obliquely from behind by north-moving plates and bounded and sliced by strike–slip faults as needed to fit, into this embayment. My contrary inference that the dominant processes were the migration of an arc (for the Carpathian arc itself) and the complex suturing of various plates, some of them far-travelled (for the inner Carpathians and intra-Carpathian region), is based on the character and distribution of subduction sutures and belts of arc-magmatic rocks, and on the lack of evidence for strike–slip faulting.

The outer part of the great arc of the Carpathian Mountains is coextensive with a belt of Tertiary thrust sheets, formed, in order concentrically inward away from the surrounding platforms, from synthrust Neogene foreland-basin strata; inner-shelf strata, Jurassic to middle Tertiary; outer-shelf strata, mostly Upper Cretaceous and Tertiary; and mostly deep-water Jurassic to middle Tertiary strata (see, for example, Unrug 1984). The inner part of the tract of deep-water materials includes much broken formation – extremely disrupted sedimentary rock (see, for example, Mastella 1988 – but otherwise the thrust belt consists of coherently deformed strata. Thrusting ceased at progressively younger time eastward, from very early Miocene in eastern Austria, progressively through the Miocene eastward across Slovakia, Poland, and Ukraine (Oszczypko & Slaczka 1989; Royden *et al.* 1983), to Pliocene at the east front of the arc in Romania.

Thrusting was radially outward around the entire arc: northwestward in western Slovakia, northward across the rest of Slovakia and Poland, northeastward in the Ukraine, eastward in central Romania, southward in southwest Romania. Deep-water materials were pushed up onto the outer continental shelf and shelf materials in turn were pushed cratonward. The thrust belt is largely a subsurface feature along the South Carpathians and is there buried by upper Neogene strata; apparently in that sector the overriding did not proceed far onto the pre-existing continental shelf.

In turn on the inner side of the belt of highly deformed deep-water strata is a concentric belt 1–20 km wide of polymict melange – chaotically disrupted and jumbled Triassic to middle Tertiary rocks, mostly sedimentary – that extends around the north limb of the arc from western Slovakia to northeast Romania (Birkenmajer 1985, 1986). The Upper Cretaceous and Palaeogene component of this melange is mostly deep-water clastic sediment, whereas older sedimentary rocks include both abyssal pelagic radiolarites and carbonate rocks and shallow-water carbonates; correlative rocks now juxtaposed occur in widely contrasted facies with quite different faunas. Crystalline rocks occurring as fragments and slices in the melange include glaucophane schist and ophiolite. Polymict melange may be lacking around the east end of the arc, although widespread broken formation, of deep-water Mesozoic through Oligocene clastic strata, is present in the same arcuate position. The equivalent belt is hidden by upper Neogene sediments along the east part of the south limb of the Carpathian arc, but along the west part polymict melange, including serpentinite-matrix melange, island-arc volcanic rocks, and continental basement rocks, reappear in the same structural position in a belt 25 km wide (Savu *et al.* 1986, 1987).

The melange of this great arc is classic subduction-complex material, and its components represent so varied a sampling of oceanic materials that it may record offscraping from thousands, not merely hundreds, of linear kilometres of subducted oceanic lithosphere. A collision, following subduction beneath an eastward-migrating arc of intervening oceanic lithosphere, with pre-existing continental shelves is indicated. The melange belt represents the unsubducted early formed part of the accretionary wedge of the advancing arc; the broken formation and deepwater sediments were accreted to the enlarging accretionary wedge as it neared the continent; the outer- and inner-shelf and foreland-basin parts of the thrust belt record the ramping of the entire wedge onto the continental shelf and the incorporation of shelf materials into the wedge. The collision was completed early in the Miocene in eastern Austria but terminated progressively later in the Miocene from Austria eastward through Slovakia, Poland, and the Ukraine, and not until the Pliocene at the east front of the arc in Romania. Benioff zone seismicity, hence slab sinking, and minor shortening still continue in Romania. I see this arc collision as due primarily to the eastward migration of a complex island arc into a remnant ocean, mostly of Jurassic lithosphere, left between continental masses incompletely sutured during late Mesozoic time. The advancing arc rode up on the surrounding continental crust progressively from west to east, like the breaking of a wave advancing into a bay, while the lengthening front of the arc continued to migrate eastward over the remaining oceanic lithosphere, with little if any strike–slip along the sides.

Behind the accretionary wedge, the arc was built partly on small continental-crustal masses, and partly was oceanic. A volcanic arc of basalt, andesite, and dacite, with subordinate rhyodacite and other rock types, trends irregularly and discontinuously along the inner part of the Carpathian arc. The volcanic rocks, like the great arc of accretionary-wedge materials, become younger eastward (Balla 1981 *a*; Poka 1988). The volcanic rocks are of early and middle Miocene age in Slovakia, middle to late Miocene or Pliocene in the Ukraine, and late Miocene and Pliocene in the east part of the arc in Romania. No volcanic-arc rocks are exposed in the east part of the south limb, but in the Apuseni Mountains, north of the west part, they are again present and are of middle and late Miocene age. The ages of volcanic rocks indicate that the arc lengthened with time as it migrated and that the east part did not yet exist when the west part was colliding with the basin margins. This behaviour is analogous to that of the Banda Arc of eastern Indonesia during Neogene time, and probably to that of the Scotia and Caribbean arcs (Hamilton 1979, 1988). Similar volcanic-arc rocks form an extensive belt, mostly subsurface, that trends west-southwestward across Hungary from the southeast corner of Slovakia; whether these are paired to the main circum-Carpathian suture, to another suture, or to both is unclear.

The analysts favouring strike–slip emplacement of a continental plate into the pre-Carpathian embayment postulate left slip along the north limb and right slip along the south limb. The radially outward thrusting around the arc, the concentric distribution and character of subduction melange, and the age and distribution of volcanic-arc rocks are among the features inexplicable in these terms.

Unlike the outer part of the Carpathian arc, the inner part is irregular and poorly defined but can be designated as a belt, 50–100 km wide, of discontinuous mountains and uplands in which structural trends tend to be concentric to the accretionary wedge. This inner belt includes most of the volcanic-arc rocks noted above, and also Palaeogene and Cretaceous magmatic-arc rocks; a nearly continuous belt of highly deformed Mesozoic and Palaeogene deep-water sediments; and a discontinuous inner tract of crustal masses of widely varied crystalline rocks (mostly late Palaeozoic but including complexes as young as Palaeogene and at least as old as early Palaeozoic) and stratigraphically overlying and tectonically interspersed upper Palaeozoic, Mesozoic, and Cenozoic sedimentary rocks. The inner Carpathian tracts trend into the better known Austrian Alps and presumably like them contain complex sutures and collision complexes. Polymict melange and blueschist are known in southeast Slovakia and northeast Hungary (Balla 1983; Faryad 1988; Reti 1988; Szadeczky-Kardoss 1973), and polymict melange is widespread in parts of central Romania (Bombita & Savu 1986; Nastaseanu 1980).

The semicircular region, 400 km in diameter, of interspersed highlands and late Neogene basins inside the Carpathian arc is not a moderately disrupted continental plate but rather is a poorly understood aggregate of accreted terranes that records Mesozoic and Cenozoic rifting, drifting, and accretion. The highlands expose widely varying Mesozoic and Palaeozoic, and possibly Precambrian, metamorphic and magmatic complexes; diverse sections of non-metamorphosed upper Palaeozoic, Mesozoic, and Palaeogene deep-water, shallow-water, and non-marine strata; and Palaeogene and Miocene volcanic rocks, mostly of arc character (Balla 1981a; Berczi-Makk 1986; Foldvary 1988; Kazmer 1986). Similar rocks, and also melange and widespread Cretaceous and Tertiary calc-alkalic arc-volcanic rocks, are known in the many wells that penetrate upper Neogene strata (Fulop & Dank 1987). The Palaeozoic and Mesozoic crystalline complexes are of types produced in convergent-plate systems but their present chaotic distribution precludes palinspastic reconstructions with available data. Much of the exposed upper Palaeozoic and upper Palaeogene sections consists of terrigenous clastic sediments, whereas the Mesozoic and lower Palaeogene are dominated by limestone and dolomite (Berczi-Makk 1986; Kazmer & Kovacs 1985), much of it in thin pelagic sections deposited on isolated submarine platforms. Facies and sections differ greatly between neighbouring crustal blocks, and even more greatly between groups of blocks, which thus likely were widely separated before aggregation. Jurassic invertebrate fossils in limestones on the crustal fragments of the northern part of the inter-arc region are of southern Mediterranean types, whereas those in the southern part are much more like those of Europe to the north: the palaeogeographic positions of the fragments not only have been reversed by complex subsequent plate motions, but the tracts likely were far apart in Mesozoic time (Burtman 1984).

Ophiolite and polymict melange, including widely varied Late Mesozoic abyssal materials, exposed in some uplands – northeast Hungary and central Romania, as noted previously; south Apuseni Mountains, Romania (Cioflica & Nicolae 1981; Savu 1984); probably southwest Hungary (Balla 1981b) – require subduction between continental crustal blocks now juxtaposed. A major belt of Tertiary melange crosses Hungary and northwest Romania. How much of the total subduction recorded is Mesozoic and how much Tertiary remains to be established.

The widespread exposed and subsurface Cretaceous and Tertiary volcanic rocks of arc types presumably were byproducts of diverse subduction systems operating between crustal blocks and island arcs now aggregated in the broad region inside the Carpathian arc (Balla 1981*a*, but cf. Balla 1986; Szadeczky-Kardoss 1973), although magmatic rocks cannot yet be confidently paired to correlative sutures with either surface or subsurface data. A mostly subsurface Miocene volcanic arc, 50 km wide, of andesites and allied rocks trends east-northeastward across all of Hungary, through the centre of the intra-Carpathian region (Balla 1981*a*; Szadeczky-Kardoss 1973). The intra-Carpathian region must record a long and complex history of rifting and convergence, including subduction of numerous oceanic plates of various sizes.

The Miocene volcanic arc noted in the preceding paragraph lies along the north side of, and may be wholly or partly paired to, a major Tertiary suture, crossing Hungary and Romania within the intra-Carpathian region, which separates the reversed-position northern and southern assemblages that were noted above. The suture was recognized as such by Szadeczky-Kardoss (1973) and forms a belt, about 50 km wide and trending west-southwestward in the subsurface of eastern Hungary, of broken formation and melange, the Senonian and Palaeogene flysch of Fulop & Dank (1987). The belt consists of highly tectonized middle Cretaceous basalt and abyssal pelagic sediments, and Upper Cretaceous through upper Oligocene marls and turbidites (Baldi-Beke *et al.* 1981; Z. Balla & T. Szederkengi, personal communication 1988; Szadeczky-Kardoss 1973). L. Lakatos translated for me descriptions of cores from four oil-exploration wells that penetrated 150–500 m into the complex. Rocks bearing Cretaceous and Palaeogene pelagic foraminifera are jumbled together with clasts containing shallow-water fossils. The descriptions are full of terms that translate as 'very dislocated', 'bright sliding surfaces', 'strong brecciation', 'shiny shear surfaces', and so on. Scaly-clay broken formation is thus widespread and polymict melange probably is extensive. I studied seismic-reflection profiles across the belt, and these show the melange as acoustic basement, devoid of coherent internal reflections, overlain by undeformed upper Miocene and younger strata. I infer formation in an accretionary wedge recording the subduction of at least hundreds of kilometres of oceanic lithosphere; and I regard as disproved the suggestion by Royden & Baldi (1988) that the complex is the Palaeogene fill of a narrow transtensional basin formed in a hypothetical strike–slip system. If the volcanic arc noted did indeed form paired to the suture, then the subduction recorded was northward beneath the northern tract. Linear anomalies shown on the magnetic map of Hungary within the obvious melange belt and west-southwest on trend from it are due primarily to Cretaceous basalt, according to sparse basement-well data as summarized schematically by Fulop & Dank 1987; I infer that the suture continues in the subsurface across southwest Hungary.

Along strike in the other direction, the suture zone crops out in north-central Romania in the 'Maramures Transcarpathian zone' (cf. Dicea *et al.* 1980; Sandulescu 1980), a belt of polymict melange, broken formation, and scaly clay that includes abundant fragments of ophiolite and abyssal sedimentary rocks, and disrupted deep-water clastic strata at least as young as Oligocene. The suture complex swings northeastward into northeast Romania and there merges with the polymict melange belt of the main Carpathian arc. I deduce that the middle Tertiary northeast-trending suture complex was in part recycled into the late Tertiary suture that there trended southeast.

Many small basins, irregular, diversely oriented, and filled by upper Neogene strata, lie in the intra-Carpathian region, primarily in Hungary (Fulop & Dank 1987) but also in each of the adjacent countries. The upper strata (the late Miocene and younger Pannonian facies) represent largely the passive filling, in general by deltas prograding from the northwest followed by lacustrine and terrestrial sedimentation, of pre-existing depressions (Mattick *et al.* 1985). The nature of most of these depressions is not established. I have looked at many published and unpublished seismic-reflection profiles across the basins. Profiles across many basins show no clear structure other than compaction in their fill. Other profiles show middle and late Miocene syndepositional shortening, for strata are continuous across asymmetric folds that define the sub-Pannonian basins, and depocentres migrate basinward going upward in the sections. Two basins – Bekes in southeast Hungary and adjacent Romania, and the southeasternmost Slovakian basin – likely mark unclosed oceanic–lithosphere 'holes', for they have positive Bouguer gravity anomalies whose amplitudes increase with increasing thickness of fill. The visualization of the Carpathian ocean gap as having been filled by a sliced continental plate in general assumes the basins to record oblique extension between linked strike–slip faults (see, for example, Royden 1988; Royden *et al.* 1983), but the assumed character of basins and faults has not been established by either outcrop or subsurface data.

The middle Tertiary suture and Miocene volcanic arc across Hungary indicate that the aggregation of the intra-Carpathian region was completed during, not before, the migration of the Carpathian arc over subducting oceanic lithosphere. It follows that the intra-Carpathian gap was broader in Palaeogene time than it is now, and that the gap was shortened from north to south by the intra-Carpathian subduction. Tertiary melange in central Romania, and middle Tertiary arc-volcanic rocks scattered about the region, presumably are also products of syn-Carpathian suturing. I see the intra-Carpathian region as produced by the squashing together of a number of small continental and island-arc fragments during Tertiary time, a view opposite to the common concept that its Tertiary history records primarily the moderate disruption of a previously coherent mass.

How many small continental and island-arc plates (terranes) are there, when and how were they aggregated, and where were they before and during aggregation? Obviously, my brief analysis raises many more questions than it answers. Evaluation is needed of the tectonic, stratigraphic, palaeoenvironmental, palaeobiogeographic, and magmatic history of each of the many possible thick-crustal masses, exposed and subsurface, in the inner-arc and intra-Carpathian regions. The region thus needs terrane analysis, which must however be integrated with actualistic plate tectonics.

REFERENCES

Baldi-Beke, M., Horvath, M. & Nagymarosy, A. 1981 *A. Rep. Hungarian geol. Inst.* **1979**, 143–158.
Balla, Z. 1981a *Earth Evolution Sci.* **1**, 240–248.
Balla, Z. 1981b *Acta mineral.-petrog. Szeged* **25**, 3–24.
Balla, Z. 1983 *A. Rep. Eotvos geophys. Inst.* **1982**, 42–65.
Balla, Z. 1986 *Tectonophysics* **127**, 213–243.
Berczi-Makk, A. 1986 *Acta geol. Hungarica* **29**, 261–282.
Birkenmajer, K. 1985 In *Geology of Poland* 1 (2) (ed. B. Slowanska & M. Bartys-Pelc), pp. 124–125, 421–443, 680–702. Warsaw: Wydawnictwa Geol.
Birkenmajer, K. 1986 *Studia geologica Polonica* **88**, 7–32.
Blake, M. C., Engebretson, D. C., Jayko, A. S. & Jones, D. L. 1985a In Howell (1985), pp. 147–157.
Blake, M. C., Jayko, A. S. & McLaughlin, R. J. 1985b In Howell (1985), pp. 159–171.
Bleahu, M. D., Boccaletti, M., Manetti, P. & Peltz, S. 1973 *J. geophys. Res.* **78**, 5025–5032.
Bombita, G. & Savu, H. 1986 *Ann. Soc. geol. Poland* **56**, 337–348.
Box, S. E. 1985 In Howell (1985), pp. 137–145.
Burchfiel, B. C. 1976 *Spec. Pap. Geol. Soc. Am.* **158**.
Burtman, V. S. 1984 *Geotectonics* **18**, 196–208.
Carlson, R. L. 1981 *Geophys. Res. Lett.* **8**, 958–961.

Cioflica, G. & Nicolae, I. 1981 *Rev. Roumaine Geol., Geophys., Geogr.: Geologie* **23**, 19–29.

Coney, P. J., Jones, D. L. & Monger, J. W. H. 1980 *Nature, Lond.* **288**, 329–333.

Csaszar, G., Haas, J., Halmai, J., Hamor, G. & Korpas, L. 1987 In *Global correlation of tectonic movements* (ed. Y. G. Leonov & V. E. Khain), pp. 173–186. London: Wiley.

Debelmas, J. 1989 In *Tectonic evolution of the Tethyan region* (ed. A. M. C. Sengor), pp. 23–42. Dordrecht: Kluwer.

Dewey, J. F., Helman, M. L., Turco, E., Hutton, D. H. W. & Knott, S. D. 1989 *Geol. Soc. Lond. Spec. Pub.* **45**, 265–283.

Dicea, O., Dutescu, P., Antonescu, F., Mitrea, G., Botez, R., Donos, I., Lungu, V. & Morosanu, I. 1980 *Romania Inst. Geol. Geofiz. Dari de Seama* **65** (4), 21–85.

Ernst, W. G., Ho, C. S. & Liou, J. G. 1985 In Howell (1985), pp. 375–389.

Faryad, S. W. 1988 *Geol. Carpathica* **39**, 747–763.

Foldvary, G. Z. 1988 *Geology of the Carpathian region*. Singapore: World Scientific.

Fulop, J. & Dank, V. (eds) 1987 *Pre-Tertiary basement geology map of Hungary*. Hungarian Geol. Inst.

Gealey, W. K. 1988 *Tectonophysics* **155**, 285–306.

Hager, B. H. & O'Connell, R. J. 1981 *J. geophys. Res.* **86**, 4843–4867.

Hamilton, W. B. 1979 *U.S. geol. Surv. prof. Pap.* **1078**.

Hamilton, W. B. 1988 *Bull. geol. Soc. Am.* **100**, 1503–1527.

Hamilton, W. B. 1989 In *Tectonic evolution of the Tethyan region* (ed. A. M. C. Sengor), pp. 655–698. Dordrecht: Kluwer.

Harper, G. D., Saleeby, J. B. & Norman, E. A. S. 1985 In Howell (1985), pp. 239–257.

Hawkins, J. W., Moore, G. F., Villamor, R., Evans, C. & Wright, E. 1985 In Howell (1985), pp. 437–463.

Hilde, T. W. C. & Uyeda, S. 1983 *Am. geophys. Union geophys. Mono.* **11**, 75–89.

Hill, L. B. 1985 In Howell (1985), pp. 173–186.

Howell, D. G. (ed.) 1985 *Tectonostratigraphic terranes of the circum-Pacific region*. Circum-Pacific Council Energy & Mineral Resources, Earth Sci., Ser. **1**.

Howell, D. G. 1989 *Tectonics of suspect terranes*. London: Chapman & Hall. (232 pages.)

Howell, D. G., Jones, D. L. & Schermer, E. R. 1985 In Howell (1985), pp. 3–30.

Irwin, W. P. 1985 In Howell (1985), pp. 187–199.

Jarrard, R. D. 1986 *Rev. Geophys.* **24**, 217–284.

Ji, X. & Coney, P. J. 1985 In Howell (1985), pp. 349–361.

Jurdy, D. M. & Stefanick, M. 1988 *J. geophys. Res.* **93**, 11,833–11,844.

Karig, D. E. 1982 *Geol. Soc. (Lond.) spec. Publ.* **10**, 563–576.

Kazmer, M. 1986 *Annls Eotvos Univ.* **26**, 45–120.

Kazmer, M. & Kovacs, S. 1985 *Acta geol. Hungarica* **28**, 71–84.

Kimbrough, D. L. 1985 In Howell (1985), pp. 285–298.

Mastella, L. 1988 *Ann. Soc. geol. Poland* **58**, 53–173.

Mattick, R. L., Rumpler, J. & Phillips, R. L. 1985 *Eotvos Geophys. Inst. geophys. Trans.* **31**, 13–54.

McCabe, R., Almasco, J. N. & Yumul, G. 1985 In Howell (1985), pp. 421–435.

Mortimer, N. 1985 In Howell (1985), pp. 201–214.

Nastaseanu, S. 1980 *Romania Inst. Geol. Geofiz. Dari de Seama* **65** (5), 109–127.

Oszczypko, N. & Slaczka, A. 1989 *Geologica Carpathica* **40**, 23–36.

Poka, T. 1988 *Mem. Am. Ass. petrol. Geol.* **45**, 257–277.

Reti, Z. 1988 *A. Rep. Hungarian geol. Inst.* **1986**, 45–52.

Royden, L. H. 1988 *Mem. Am. Ass. petrol. Geol.* **45**, 27–48.

Royden, L. H. & Baldi, T. 1988 *Mem. Am Ass. petrol. Geol.* **45**, 1–16.

Royden, L. H. & Burchfiel, B. C. 1989 *Tectonics* **8**, 51–61.

Royden, L. H., Horvath, F. & Rumpler, J. 1983 *Tectonics* **2**, 63–90.

Sandulescu, M. 1980 *Romania Inst. Geol. Geofiz. Dari de Seama* **65** (5), 163–180.

Sandulescu, M. 1988 *Mem. Am. Ass. petrol. Geol.* **45**, 17–25.

Savu, H. 1984 *Rev. Roumaine Geol., Geophys., Geog.: Geologie* **29**, 36–43.

Savu, H., Udrescu, C. & Neacsu, V. 1985 *Rev. Roumaine Geol., Geophys., Geog.: Geologie* **29**, 45–53.

Savu, H. *et al.* 1987 *Rev. Roumaine Geol., Geophys., Geog.: Geologie* **31**, 19–27.

Szadeczky-Kardoss, E. 1973 *Bull. Hungarian geol. Soc.* **103**, 224–244.

Taylor, B. & Karner, G. D. 1983 *Rev. Geophys.* **21**, 1727–1741.

Unrug, R. 1984 *Ann. Soc. geol. Poland* **52**, 39–66.10

Discussion

P. D. CLIFT (*Grant Institute of Geology, University of Edinburgh, U.K.*). Tectonic reconstructions of active continental margins involving large-scale, along strike motion of accreted continental and oceanic terranes may have only a limited application to narrow ocean basins, such as those whose closure formed the European Alpine system. The absence of island-arc terranes from suture zones any further west than the Aegean Sea, together with palaeomagnetic data from the African and Eurasian plates constrains the western Tethys to being a relatively small embayment of a larger easterly ocean, and experiencing only limited along strike motion due to different times of opening of the central and north Atlantic Ocean. Continental terrane motion was thus broadly a north to south migration. Despite the small size of the oceans, the rates of continental rifting and oceanic subduction, along the margin of Neotethys, were, however, comparable with modern circum-Pacific examples, although in the Tethyan case punctuated by long period of relative tectonic quiescence.

W. B. HAMILTON. Although Dr Clift is of course correct that western Tethys was a narrow ocean, I disagree with his derivative assumptions. He suggests that motions of intra-Tethyan plates were primarily meridional, but Italy has moved eastward over a subducting Adriatic plate (Dewey *et al.* 1989), Corsica and Sardinia also have migrated primarily eastward, and the advancing arc that collided with the Spanish and African margins of what is now the Alboran Sea must have had a large westward component of motion. My analysis of the Carpathian arc as eastward-migrating is given in the preceding text.

Nor are island arcs lacking west of the Aegean Sea; magmatic-arc rocks, variously oceanic, continental, and hybrid, are voluminous in the Dinarides, Carpathians, and Italy, and present in lesser quantities in other parts of the western regions. Some western Tethyan complexes may well record major subduction.

A. H. F. ROBERTSON (*Grant Institute of Geology, University of Edinburgh, U.K.*). Palaeomagnetic evidence appears to suggest that much of southern Greece and Turkey (at least) originated near the northern margin of Gondwanaland rather than far to the east in the ancestral Pacific ocean. Other units, especially those located along the northern margins of the Tethys (e.g. Pontides) may be much more exotic, however. The implication that plate displacement rates were unusually slow is perhaps misleading. At least along the North Gondwana margin relatively rapid events (e.g. Triassic spreading in several small ocean basins) were separated by long periods of relative quiescence (e.g. with development of passive margins). Rapid displacements were often episodic, for example, as a consequence of the motion of larger plates (e.g. Late Jurassic opening of the North Atlantic, Cretaceous opening of the South Atlantic).

W. B. HAMILTON. Dr Robertson, like Dr Clift, infers that intra-Tethyan motions were episodic, periods of rapid plate migrations having been separated by long periods of stability. Although long-continuing passivity of specific continental margins or small plates is indeed obvious, other western Tethyan plates and margins were undergoing deformation and magmatism during at least most of Mesozoic and Cenozoic time. Such contrasts – passive here, subducting there – necessarily typify all regions of moving plates, and are not indications of episodicity.

Palaeomagnetic evidence, as referred to by Dr Robertson, can not constrain palaeolongitude, and so can provide at best only part of the data needed to cope with the problems of an ocean that was elongate east–west. Palaeontologic evaluation is needed to define biogeographic provinciality and to explore the possibility of large offsets of displaced materials within the vast Tethyan complexes.

Geological constraints on the origin of the mantle root beneath the Canadian shield

By P. F. Hoffman

Geological Survey of Canada, 601 *Booth Street, Ottawa, Ontario, Canada K1A 0E8*

Cratonic North America is composed of a cluster of Archaean microcontinents centred on the Canadian shield, and juvenile Proterozoic crust that lies mainly buried beneath the sedimentary cover of the western and southern interior platforms. The shield is underlain by an anomalous low-temperature mantle root that is absent beneath the platform. As there appears to be no systematic difference in crustal thickness or density between the shield and the platform, the long-lived arching of the shield implies an intrinsic buoyancy imparted by the mantle root that more than offsets its colder temperature. Isotopic and seismic anisotropy data indicate an Archaean age for the mantle root, close to the time of formation of the overlying crust. The preferential development of the mantle root beneath Archaean crust is consistent with an origin by imbrication of partly subducted slabs of highly depleted oceanic lithosphere, assuming that buoyant subduction was more common in the Archaean. Formation of the mantle root was not dependent on collisional orogenesis, as has been suggested, but the Archaean cratonic mantle was sufficiently buoyant and refractory to survive later tectonic thickening. The mantle root persists beneath Archaean crust that was transected by mafic dyke swarms and subjected to short-lived episodes of post-orogenic crustal melting, but the root is reduced at mantle plume initiation sites. The partitioning of Archaean and Proterozoic crust between the shield and the platform, respectively, causes the shield to misrepresent Precambrian crust as a whole. Studies of the shield falsely conclude that a high percentage of Precambrian crust formed in the Archaean, and that the Proterozoic was characterized by epicontinental volcanism and sedimentation, and crustal 'reworking'. Furthermore, the isotopic ratios of detritus eroded from the craton may tend to overestimate the mean age of continental crust.

1. Introduction

Shields are arched areas of continental crust that expose Precambrian igneous and metamorphic rocks. Areas where Precambrian crystalline rocks are hidden by little-deformed sedimentary or volcanic cover are called continental *platforms*. *Cratons* encompass shields and contiguous platforms.

For obvious reasons, shield areas provide most of our geological knowledge of Precambrian crust. The largest shield, the Canadian shield (figure 1), is composed 84% of crust that separated from the mantle before 2.5 Ga (McCulloch & Wasserburg 1978; Patchett & Arndt 1986; Hoffman 1989; E. Hegner, personal communication 1989). This fact has contributed to the notion that most Precambrian crust originated in the Archaean (4.0–2.5 Ga), and that Proterozoic (2.5–0.57 Ga) tectonism was dominated by 'reworking' of Archaean crust, and by epicontinental sedimentation and volcanism (see, for example, Windley 1984). The validity of this generalization depends on the assumption that shields are representative of all Precambrian crust.

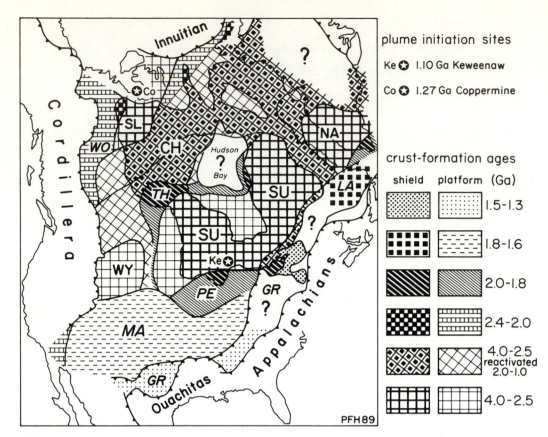

FIGURE 1. Crust-formation age map of the North American craton, based on Sm–Nd, Pb–Pb and U–Pb geochronology of samples from the shield and the subsurface (data from McCulloch & Wasserburg 1978; Nelson & DePaolo 1985; Patchett & Arndt 1986; Bennett & DePaolo 1987; Patchett & Ruiz 1989; E. Hegner, personal communication 1989; and other sources cited in Hoffman (1988, 1989)). Note the concentration of Archaean crust in the shield. Abbreviations: CH, Churchill hinterland; *GR*, Grenville orogen; *LA*, Labradorian orogen; *MA*, Mazatzal orogen; *NA*, Nain province; *PE*, Penokean orogen; SL, Slave province; SU, Superior province; *TH*, Trans-Hudson orogen; *WO*, Wopmay orogen; WY, Wyoming province. Breaks in pattern indicate possible cryptic sutures within the Churchill hinterland. Greenland is positioned in a pre-drift reconstruction after Roest & Srivastava (1989).

Recent work shows that, unlike the Canadian shield, the North American platform contains a high proportion (59%) of juvenile Proterozoic crust (figure 1). This conclusion is based on Sm–Nd model mantle-separation ages (De Paolo 1981), amplifying earlier Rb–Sr and Pb–Pb data, of Precambrian inliers and scores of basement samples obtained by commercial drilling through Phanerozoic sedimentary cover (Nelson & DePaolo 1985; Bennett & DePaolo 1987; Patchett & Ruiz 1989). The Canadian shield is therefore not representative of the North American craton as a whole, but is strongly biased in favour of Archaean crust.

Maps of isopachs and lithofacies of sedimentary cover on the North American platform (Cook & Bally 1975) indicate that the Canadian shield has been an area of relatively elevated basement throughout the Phanerozoic. Presumably this reflects long-lived lateral heterogeneities in the crust and/or mantle which migrate with the lithospheric plate. In this regard, the Canadian shield is distinct from shields produced by active tectonics. Uplift of the Arabian–Nubian shield, for example, is probably a consequence of Neogene rifting in the Red Sea. Uplift due to the thermal and dynamic effects of rifting will be geologically short-lived, however, once rifting ceases. The only permanent component of uplift adjacent to the rift zone would be that due to subsurface addition of basaltic melts to the crust (cf. White & McKenzie 1989).

Relative to the rest of the North American craton, the Canadian shield is not an area of unusually thick crust but coincides with a region of anomalous upper mantle. Crustal thickness appears not to vary systematically between the shield and platform (Mooney & Braile 1989), so that relative uplift of the shield is not simply an isostatic consequence of a thicker crust. The shield is, however, underlain by an anomalous mantle 'root', which is indicated by seismic shear velocities (figure 2) that are faster than those beneath the platform to depths exceeding 200 km (Grand 1987). The higher velocities signify colder temperatures within the mantle root relative to the adjacent asthenosphere, the implications of which are discussed in detail by Jordan (1988). To resist convective erosion, the mantle root must be more refractory than the adjacent asthenosphere. To ensure gravitational stability, the root must be composed of intrinsically less dense material to offset its lower temperature. Otherwise, the formation of a cold mantle root would result in subsidence equivalent to a sedimentary basin over 15 km deep or, alternatively, would generate a large positive gravity anomaly, neither of which are observed on the scale of the shield. The twin requirements of intrinsic buoyancy and elevated solidus temperature are met by mantle material that is highly depleted in the components extracted by partial melting. The extraction of basaltic melt depletes the mantle residuum in iron, thereby decreasing its density by reducing its capacity to form garnet. The fact that a depleted mantle root (the *tectosphere* of Jordan 1988) coincides roughly with the area of the

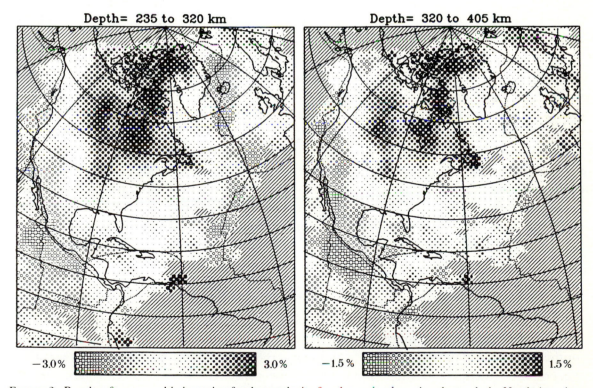

FIGURE 2. Results of tomographic inversion for shear velocity for planar depth sections beneath the North American plate according to Grand (1987). Hatching indicates unresolved areas. Note that the velocity perturbation scale changes as a function of depth. Compare the location of the high-velocity mantle root with the area of the shield in figure 1. Note the present-day position of Greenland, unlike that in figure 1.

Canadian shield (Grand 1987) implies that the mantle root has a net buoyancy and is responsible for the arching of the crust.

Different causal mechanisms for cratonic mantle roots have been advanced. Jordan (1988) proposes a two-stage process. First, subcontinental mantle is depleted by partial melting at an Andean-type magmatic arc. Then, it is advectively thickened during collisional orogenesis. This model implies that depleted mantle roots should preferentially underlie crustal regions that have experienced arc magmatism and subsequent collisional orogenesis. Conversely, Ashwal & Burke (1989) argue that collisional orogenesis leads to delamination or convective erosion of the tectonically thickened mantle lithosphere, whereupon it is replaced by fertile asthenospheric mantle (cf. England & Houseman 1988). They view continents as aggregates of island arcs having roots of depleted mantle that are preserved only in regions that have not experienced subsequent collisional orogenesis. Neither Jordan (1988) nor Ashwal & Burke (1989) predict a secular control on the formation of mantle roots.

In contrast, Helmstaedt & Schulze (1989) envision cratonic mantle roots as being formed of imbricated slabs of partly subducted oceanic lithosphere. Their model, which is based on the study of xenoliths of Archaean age contained in kimberlites intruding the Kaapvaal craton of southern Africa, is particularly relevant to the Archaean. It presupposes a buoyant mode of subduction, which would have been more common at that time. Oceanic lithosphere entering Archaean subduction zones would have been relatively young and warm because of higher global spreading rates (Bickle 1986). Furthermore, the oceanic crust would have been relatively thick and the suboceanic mantle strongly depleted because of the large volumes of melt produced at Archaean spreading ridges (Bickle 1986). Density-driven subduction of Archaean oceanic slabs would have been dependent on the formation of eclogite within the oceanic crust.

The association of mantle roots and Archaean crust may also be a result of selective preservation. Presuming a secular decline in mean mantle temperature, the Archaean mantle would have convected rapidly due to its relatively low viscosity. Richter (1988) suggests that mantle convection was so vigorous as to cause recycling of Archaean crust unprotected by refractory mantle roots.

In summary, different genetic models predict that the tectosphere should occur either in regions that have experienced collisional orogenesis (Jordan 1988), regions that have escaped collisional reactivation (Ashwal & Burke 1989), or regions that have been mechanically underplated by partly subducted oceanic slabs (Helmstaedt & Schulze 1989). The purpose of this paper is to evaluate these models in light of the distribution of the mantle root with respect to segments of the North American craton having different tectonic histories.

2. Extent of the mantle root beneath the North American craton

Cratonic North America is composed of a cluster of Archaean microcontinents, centred on the Canadian shield, and juvenile Proterozoic crust that mainly underlies the western and southern interior platforms (figure 1). Hoffman (1988, 1989) documents that aggregation of the craton in detail and those works should be consulted for references concerning the geological and geochronological information outlined below. Aggregation of the Archaean microcontinents occurred in a series of collisions between 1.97 and 1.82 Ga. Proterozoic arc terrances built on crust having mantle-separation ages of 2.4–2.0 Ga were accreted to the western margin (Wopmay orogen) of the Archaean protocraton between 1.90 and about 1.8 Ga. Juvenile (i.e. newly extracted from the mantle) Proterozoic crust was accreted to the southern margins of the Superior and Nain provinces (Penokean and Ketilidian orogens, respectively) between 1.90 and 1.80 Ga. More juvenile crust was accreted to the southern margin of the protocraton (Mazatzal and Labradorian orogens) between 1.80 and 1.60 Ga. Finally, crust having mantle-separation ages of 1.5–1.3 Ga was accreted in parts of the Grenville orogen between 1.2 and 1.0 Ga.

The ratio of Archaean to Proterozoic crust is almost 6:1 for the shield but only 0.66:1 for the platform. Exceptions to the general association of Archaean crust with the shield and Proterozoic crust with the platform may be short-lived. They include the Proterozoic crust of the Laurentian highlands (Grenville orogen) adjacent to the St Lawrence rift system, and Cretaceous–Tertiary downwarping of Archaean crust in the foreland of the Rocky Mountains in west-central North America (Mitrovica *et al.* 1989).

The Superior and Slave provinces (figure 1) are composed mostly of 3.0–2.7 Ga island arcs and relatively small fragments of older continental crust. Amalgamation of the island arcs of the Superior province occurred between 2.73 and 2.70 Ga, and was immediately followed by widespread crustal melting between 2.69 and 2.67 Ga. The Slave province experienced a broadly similar scenario that was delayed by about 50 Ma relative to the Superior province. Significant post-Archaean tectonic shortening is limited to within 100 km of the margins of both provinces, except for a WNS-dipping intracratonic thrust in the central Superior Province (Percival *et al.* 1989). Both provinces are transected by Proterozoic mafic dyke swarms (Fahrig & West 1986) but were spared the mid-Proterozoic (1.8–1.1 Ga) crustal melting that affected other parts of the North American craton.

High-velocity mantle roots underlie both the Slave and Superior provinces, with the latter having the highest velocities and deepest root in North America (figure 2). The difference in velocity structure may reflect the small area of the Slave province, which approaches the horizontal resolving power (400 km) of the tomographic inversion (Grand 1987). If real, the difference might signify the partial destruction of the mantle root beneath the Slave province during an episode of subduction (1.88–1.86 Ga) along its western margin. There is no geological evidence for subduction beneath the Superior province.

Seismic anisotropy data provide a potential key to the age of mantle roots. In at least one area of the Superior province, vertically travelling seismic waves have an ENE azimuth of fast polarization that parallels the island arcs within the province and lies within the bulk flattening plane developed during their amalgamation (Silver & Chan 1988). The seismic anisotropy presumably reflects a strain fabric within the mantle root and the magnitude of the anisotropy implies that the strain fabric likely persists through a depth range on the order of 200 km (Silver & Chan 1988). The anisotropy data, although limited in areal extent, are none the less important because, if confirmed, they imply that the mantle root formed in the Archaean. Certainly, the province underwent no significant bulk strain after intrusion of the NNW-trending Matachewan dyke swarm at 2.45 Ga. The inferred age of the mantle root accords with evidence from diamond inclusions, xenocrysts and mantle xenoliths contained in kimberlites intruding the Kaapvaal craton, which unequivocally indicate an Archaean age for the tectosphere in southern Africa (Boyd & Gurney 1986). The persistence of a depleted mantle lithosphere of Archaean age beneath the Superior province is also supported by Nd and Sr initial isotopic ratios in carbonatite intrusions ranging in age from 2.7 to 0.11 Ga (Bell &

Blenkinsop 1987). The apparent Archaean age of the mantle root is compatible with the model of Helmstaedt & Schulze (1989), although they do not explicitly address the problem of the origin of the strain fabric responsible for the seismic anisotropy.

Mantle roots are conspicuously absent beneath the Archaean Wyoming and Nain provinces (figures 1 and 2). Both contain early–middle Archaean gneiss terranes that were thoroughly reactivated when each province was amalgamated in the late Archaean. This accords with the Ashwal & Burke (1989) model, in which mantle roots are destroyed by collisional orogeny. Alternatively, destruction of a mantle root beneath the Wyoming province may be related to Laramide horizontal subduction and crustal shortening (Bird 1988). Moreover, the Iceland plume may have eroded a former mantle root beneath the Nain province during rifting of the northern North Atlantic (White & McKenzie 1989). Accordingly, the absence of mantle roots beneath the Wyoming and Nain provinces may be related to Mesozoic–Cenozoic tectonism rather than Archaean collisional orogenesis.

The middle–late Archean provinces of the Churchill hinterland (the Rae, Hearne and Burwell provinces of Hoffman (1989), figure 1) were variably reactivated as a result of collisions with the Slave, Superior and Nain provinces between 1.97 and 1.82 Ga. As a result of crustal shortening, early Proterozoic sedimentary cover is preserved in cuspate synclinoria and Klippen between broad basement antiforms. West of Hudson Bay, crustal shortening in the Churchill hinterland was succeeded by extensive alkalic magmatism and associated normal faulting at about 1.85 Ga, widespread crustal melting (rapakivi-type granite and rhyolite) at 1.76 Ga, and development of cratonic basins about 1.7 Ga. On southern Baffin Island, northeast of Hudson Bay, an enormous charnockite-granite batholith, emplaced at 1.90–1.85 Ga, is flanked by high-grade early Proterozoic fold belts. Nevertheless, both regions are underlain by high-velocity mantle roots (figure 2). An Archaean mantle root under the Churchill hinterland may have been sufficiently refractory and buoyant to survive tectonic thickening, contrary to the Ashwal & Burke (1989) model, but how it could have survived advective heating sufficient to promote extensive crustal melting is problematic. Sm–Nd model ages of the 1.85 Ga alkaline rocks west of Hudson Bay are most compatible with melting of an Archaean cratonic mantle source (Esperança & LeCheminant 1986), but isotopic evidence bearing on the age of the mantle root that survived the 1.76 Ga crustal melting event is lacking. Seismic anisotropy data might potentially resolve the problem of the age of the extant mantle root west of Hudson Bay. If the root post-dates the crustal melting event, it should have no azimuthal anisotropy because the resulting granite–rhyolite suite remains undeformed. On the other hand, if the root predates the crustal melting event, an azimuthal anisotropy related to the late Archaean and early Proterozoic NW–SE shortening events would be expected.

Mantle roots are not strictly limited to Archaean cratons, as implied by Richter (1988) and Hawkesworth *et al.* (1990). The Penokean, Trans-Hudson and Wopmay orogens (figure 1) incorporate crust that was extracted from the mantle in the early Proterozoic. In the Wopmay orogen, 1.95–1.85 Ga magmatic arcs, built on 2.3–2.0 Ga crust, were accreted to the Archaean Slave province between 1.90 and about 1.8 Ga. In the Trans-Hudson orogen, juvenile 1.91–1.85 Ga island arcs are sandwiched between the Archaean Superior and Hearne provinces southwest of Hudson Bay. In the Penokean orogen, 1.89–1.84 Ga island arcs, in part built on Archaean crustal slivers, were accreted to the Superior province between 1.85 and 1.84 Ga. All three orogens are underlain by high-velocity mantle roots, although the roots are less well developed than beneath the adjacent Archaean provinces (figure 2).

Sites of profuse basaltic volcanism associated with the initiation of mantle plumes (cf. Richards *et al.* 1989) correlate with areas of reduced mantle shear velocities beneath Archaean crust, apparently signifying significant erosion of the mantle root (figure 2). This is observed in the northern Slave province (figure 1), the site of a 1.27 Ga mantle plume that was responsible for the Coppermine plateau basalts and the enormous Mackenzie dyke swarm (LeCheminant & Heaman 1989). The Keweenaw (Midcontinent) rift coincides with a corridor of lower-velocity mantle between the Penokean orogen and the Superior province (figure 1). The rift, which overlies a mafic crustal underplate and contains up to 20 km of basalt erupted between 1.10 and 1.09 Ga, developed above a mantle plume (Nicholson *et al.* 1989). Near the Grenville orogen, the NW–SE trend of the rift parallels the direction of contemporaneous thrusting in the orogen, consistent with the 'impactogen' model of Burke (1980). Away from the orogen, the rift-trend is deflected southwestward, tracking the Superior–Penokean boundary zone, implying that the rift was unable to propogate into the rigid interior of the Superior province. The lower-velocity corridor beneath the rift is compatible with the suggestion that channels of relatively 'fertile' mantle may replace depleted mantle lithosphere during intracratonic rifting (Phipps 1988). Erosion of the Archaean mantle root by the Keweenaw plume may be compared with the postulated destruction of an Archaean mantle root beneath the Nain province by the Iceland plume.

Mantle roots are absent or poorly developed beneath the Proterozoic crust on the southern and southeastern margins of the craton (figures 1 and 2). The Mazatzal orogen is composed of 1.80–1.70 Ga juvenile crust that was reactivated by NNW-directed thrusting between 1.69 and 1.65 Ga. Protracted anorogenic magmatism including widespread crustal melting occurred between 1.54 and 1.32 Ga. Proterozoic crust in the southwestern United States was reactivated again during Neogene Basin-and-Range extension. The Grenville orogen experienced northwest-directed crustal-scale thrusting between 1.2 and 1.0 Ga. South of the Superior province, Grenvillian thrusting involved parautochthonous Archaean and Penokean (*ca.* 1.9 Ga) crust, and allochthonous terranes having 1.5–1.3 Ga model ages (figure 1). Near the Atlantic coast, Grenvillian thrusting involved mainly of 1.70–1.65 Ga (Labradorian) crust. Both the Labradorian and the Penokean zones of the Grenville orogen were intruded episodically by associated anorthositic and granitic batholiths between 1.65 and 1.15 Ga. Mantle shear velocities beneath the Grenville orogen are higher in the area of the Canadian shield than the buried platform, consistent with the association of the mantle root with crustal arching, but the reason for the difference in mantle structure cannot be assessed until the buried part of the Grenville orogen is better known.

3. CONCLUSIONS AND DISCUSSION

The Canadian shield is not representative of the North American craton as a whole. Crust that evolved from the mantle in the Archaean (before 2.5 Ga) makes up 84% of the shield but only 55% of the entire craton. The shield gives a misleading impression that little new crust evolved from the mantle during the Proterozoic.

The area of the Canadian shield coincides closely with the area underlain by a high-velocity mantle root, or tectosphere, that extends to depths of at least 200 km. The mantle root must be composed of relatively refractory material of low intrinsic density to offset its lower temperature. The arching of the shield, which was located in the same general region throughout the Phanerozoic, may result from a net buoyancy of the mantle root.

The mantle root is not limited to areas of Archaean crust but is best developed there. It is less well developed beneath crust extracted from the mantle in the early Proterozoic, and least well developed beneath Middle Proterozoic and younger crust. The secular control on the development of the mantle root is compatible with the Helmstaedt & Schulze (1989) model, assuming that buoyant subduction has become less common over geologic time. If the arching of the shield is caused by a buoyant mantle root developed preferentially beneath Archaean crust, this would account for the observed partitioning of Archaean and Proterozoic crust between the shield and the platform respectively.

Isotopic and seismic anisotropy data suggest that the mantle root beneath the Archaean provinces formed in the Archaean, close to the time of crust formation, rather than developing progressively over subsequent geologic time.

Development of the mantle root is not dependent on collisional orogenesis, as implied by Jordan (1988). The mantle root is best developed under the Archaean Superior province, which is composed of island arcs that underwent little tectonic shortening after their initial aggregation.

On the other hand, collisional orogenesis does not always destroy a mantle root, as inferred by Ashwal & Burke (1989). A mantle root persists beneath the Archaean crust of the Churchill hinterland, which was reactivated during early Proterozoic collisions with the Slave, Superior and Nain microcontinents. Yet, mantle roots are conspicuously absent beneath the Wyoming and Nain provinces, which experienced late Archaean collisional orogenesis. However, destruction of their mantle roots may have occurred in the Mesozoic–Cenozoic, as a consequence of the Laramide orogeny in the Wyoming province and the Iceland plume in the Nain province.

The efficacy of mantle plume initiation in eroding mantle roots beneath Archaean crust is also shown by areas of reduced shear velocity near the sites of the Coppermine plume (1.27 Ga) in the northern Slave province and the Keewanaw plume (1.10 Ga) in the southern Superior province.

Mantle roots beneath Archaean crust survived the emplacement of extensive mafic dyke swarms, and also crustal underplating by mafic melts presumed responsible for short-lived post-orogenic crustal melting events (e.g. 1.76 Ga rapakivi-type granite and rhyolite suites in the Churchill hinterland west of Hudson Bay). However, little or no mantle root is present beneath the early Proterozoic crust of the southern United States, which was subjected to repeated anorogenic melting between 1.5 and 1.3 Ga.

The Archaean crust of the shield has been exposed to erosion for much of Phanerozoic time. The Proterozoic crust of the platform has been largely protected from erosion by sedimentary cover since the early Palaeozoic. As a result, the Archaean crust has contributed disproportionally to the detritus eroded from the North American craton, including modern river sediments. Consequently, the mean age of continental crust, which is used to constrain models of chemical cycling between the crust and mantle (see, for example, Galer et al. 1989), will tend to be overestimated by the isotopic ratios of particulate and dissolved material in rivers draining the North American craton (cf. Goldstein et al. 1984; Goldstein & Jacobsen 1987). This should also be true of ancient sediments of cratonic derivation deposited after the early Palaeozoic transgression of the platform.

MANTLE ROOT

If mantle roots developed preferentially beneath Archaean crust are responsible for long-lived shields, then the spatial association of shields and Archaean crust should be observed on continents other than North America. The largest shield area in Australia (Western Australian shield) is composed almost entirely of Archaean crust. Except for the Archaean Gawler craton of southern Australia, most of the rest of the pre-Palaeozoic craton of Australia is composed of crust having Proterozoic model ages (McCulloch 1987). For other continents, crust-mantle extraction ages are poorly known, especially in platform areas, and upper mantle shear velocity structures remain to be worked out.

I thank Stephen Grand for permission to reproduce figure 2, Chris Hawkesworth and Doug Nelson for relevant preprints, Dallas Abbott, Tom Jordan and Paul Silver for discussions, and Abbott, Hawkesworth and Stephen Lucas for helpful comments on a draft of the manuscript. This is Geological Survey of Canada contribution 45089.

References

Ashwal, L. D. & Burke, K. 1989 African lithospheric structure, volcanism, and topography. *Earth planet. Sci. Lett.* **96**, 8–14.

Bell, K. & Blenkinsop, J. 1987 Archean depleted mantle: evidence from Nd and Sr initial isotopic ratios of carbonatites. *Geochim. Cosmochim. Acta* **51**, 291–298.

Bennett, V. C. & DePaolo, D. J. 1987 Proterozoic crustal history of the western United States as determined by neodymium isotopic mapping. *Bull. geol. Soc. Am.* **99**, 674–685.

Burke, K. 1980 Intracontinental rifts and aulacogens. In *Continental tectonics, studies in geophyusics* (ed. B. C. Burchfiel and others), pp. 42–49. Washington, D.C.: American Geophysical Union.

Bickle, M. J. 1986 Implications of melting for stabilization for the lithosphere and heat loss in the Archaean. *Earth Planet. Sci. Lett.* **80**, 314–324.

Bird, P. 1988 Formation of the Rocky Mountains, western United States: a continuum computer model. *Science, Wash.* **239**, 1501–1507.

Boyd, F. R. & Gurney, J. J. 1986 Diamonds and the African lithosphere. *Science, Wash.* **232**, 472–477.

Cook, T. D. & Bally, A. W. 1975 *Stratigraphic atlas of North and Central America.* Princeton University Press. (272 pages.)

DePaolo, D. J. 1981 Neodymium isotopes in the Colorado Front Range and crust-mantle evolution in the Proterozoic. *Nature, Lond.* **291**, 193–196.

England, P. C. & Houseman, G. A. 1988 The mechanics of the Tibetan plateau. *Phil. Trans. R. Soc. Lond.* A **326**, 301–320.

Esperança, S. & LeCheminant, A. N. 1986 Isotopic evidence for multiple enrichment events from mica-lamprophyre dykes in the District of Keewatin, Canada. *Geol. Soc. Am. Abst. Programs* **18**, 595.

Fahrig, W. F. & West, T. D. 1986 *Diabase dyke swarms of the Canadian shield.* Geological Survey of Canada Map 1627A, scale 1:4873900.

Galer, S. J. G., Goldstein, S. L. & O'Nions, R. K. 1989 Limits on chemical and convective isolation in the Earth's interior. *Earth planet. Sci. Lett.* **75**, 257–290.

Goldstein, S. J. & Jacobsen, S. B. 1987 The Nd and Sr isotopic systematics of riverwater dissolved material: implications for the sources of Nd and Sr in seawater. *Chem. Geol. (Isotope Geosci. Section)* **66**, 245–272.

Goldstein, S. L., O'Nions, R. K. & Hamilton, P. J. 1984 A Sm–Nd isotopic study of atmospheric dusts and particulates from major river systems. *Earth planet. Sci. Lett.* **70**, 221–236.

Grand, S. P. 1987 Tomographic inversion for shear velocity beneath the North American plate. *J. geophys. Res.* **92**, 14065–14090.

Hawkesworth, C. J., Kempton, P. D., Rogers, N. W., Ellam, R. M. & van Calsteren, P. W. 1990 Continental mantle lithosphere, and shallow level enrichment processes in the Earth's mantle. *Earth planet. Sci. Lett.* **96**, 256–268.

Helmstaedt, H. & Schulze, D. J. 1989 Southern African kimberlites and their mantle sample: implications for Archaean tectonics and lithosphere evolution. In *Kimberlites and related rocks, vol. 1. Their composition, occurrence, origin, and emplacement* (ed. J. Ross). *Geol. Soc. Australia, Spec. Publ.* **14**, pp. 358–368. Carlton, Australia: Blackwell.

Hoffman, P. F. 1988 United plates of America, the birth of a craton: Early Proterozoic assembly and growth of Laurentia. *A. Rev. Earth planet. Sci.* **16**, 543–603.

Hoffman, P. F. 1989 Precambrian geology and tectonic history of North America. In *The geology of North America – an overview* (ed. A. W. Bally & A. R. Palmer). *Geol. Soc. Am., The geology of North America vol. A,* pp. 447–512.

Jordan, T. H. 1988 Structure and formation of the continental tectosphere. *J. Petrol., Spec. Lithosphere Issue,* pp. 11–37.

LeCheminant, A. N. & Heaman, L. M. 1989 Mackenzie igneous events, Canada: Middle Proterozoic hotspot magmatism associated with ocean opening. *Earth planet. Sci. Lett.* **96,** 38–48.

McCulloch, M. T. 1987 Sm–Nd isotopic constraints on the evolution of Precambrian crust in the Australia continent. In *Proterozoic lithospheric evolution* (ed. A. Kröner). *Am. Geophys. Union, Geodynamics Ser.* **17,** 115–130.

McCulloch, M. T. & Wasserburg, G. J. 1978 Sm–Nd and Rb–Sr chronology of continental crust formation. *Science, Wash.* **200,** 1003–1011.

Mitrovica, J. X., Beaumont, C. & Jarvis, G. T. 1989 Tilting of continental interiors by the dynamic effects of subduction. *Tectonics* **8,** 1079–1094.

Mooney, W. D. & Braile, L. W. 1989 The seismic structure of the continental crust and upper mantle of North America. In *The geology of North America – an overview* (ed. A. W. Bally & A. R. Palmer). *Geol. Soc. Am., The geology of North America vol. A,* pp. 39–52.

Nelson, B. K. & DePaolo, D. J. 1985 Rapid production of continental crust 1.7 to 1.9 b.y. ago: Nd isotopic evidence from the basement of the North American mid-continent. *Bull. geol. Soc. Am.* **96,** 746–754.

Nicholson, S. W., Shirey, S. B. & Schulz, K. J. 1989 1100-Ma Keweenaw hotspot: Nd and Pb isotopic evidence for a Proterozoic mantle plume in the Midcontinent rift, U.S.A. *Eos* **70,** 1357.

Patchett, P. J. & Arndt, N. T. 1986 Nd isotopes and tectonics of 1.9–1.7 Ga crustal genesis. *Earth planet Sci. Lett.* **78,** 329–338.

Patchett, P. J. & Ruiz, J. 1989 Nd isotopes and the origin of Grenville-age rocks in Texas: implications for Proterozoic evolution of the United States mid-continent region. *J. Geol.* **97,** 685–695.

Percival, J. A., Green, A. G., Milkereit, B., Cook, F. A., Geis, W. & West, G. F. 1989 Seismic reflection profiles across deep continental crust exposed in the Kapuskasing uplift. *Nature, Lond.* **342,** 416–420.

Phipps, S. P. 1988 Deep rifts as sources for alkaline intraplate magmatism in eastern North America. *Science, Wash.* **334,** 27–31.

Richards, M. A., Duncan, R. A. & Courtillot, V. E. 1989 Flood basalts and hot-spot tracks: plume heads and tails. *Science, Wash.* **246,** 103–107.

Richter, F. M. 1988 A major change in the thermal state of the earth at the Archean–Proterozoic boundary: consequences for the nature and preservation of continental lithosphere. *J. Petrol., Spec. Lithosphere Issue,* pp. 39–52.

Roest, W. R. & Srivastava, S. P. 1989 Sea-floor spreading in the Labrador Sea: a new reconstruction. *Geology* **17,** 1000–1003.

Silver, P. G. & Chan, W. W. 1988 Implications for continental structure and evolution from seismic anisotropy. *Nature, Lond.* **335,** 34–39.

White, R. & McKenzie, D. 1989 Magmatism at rift zones: the generation of volcanic continental margins and flood basalts. *J. geophys. Res.* **94,** 7685–7729.

Windley, B. F. 1984 *The evolving continents,* 2nd edn. New York: Wiley. (399 pages.)

A geochemical approach to allochthonous terranes: a Pan-African case study

By N. B. W. Harris, I. G. Gass, F.R.S., and C. J. Hawkesworth

Department of Earth Sciences, Open University, Milton Keynes MK7 6AA, U.K.

The recognition of Mesozoic and Cenozoic terranes can best be made from palaeomagnetic, structural and palaeontological studies, but older regions of continental crust require geochemical constraints to evaluate crustal growth through terrane accretion. For Precambrian shields, the pattern of Pb and Nd isotopic provinces may reveal the mechanism of crustal growth.

The Afro-Arabian Shield was generated by calc-alkaline magmatism between 900 and 600 Ma ago. This example of Pan-African crustal growth underlies an area of at least 1.2×10^6 km², which may extend to 3.5×10^6 km² beneath Phanerozoic sediments and Tertiary volcanic cover. Field evidence and trace element geochemistry suggest that Pan-African tectonics began as a series of intra-oceanic island arcs that were accreted to form continental lithosphere over a period of 300 Ma. The great majority of Nd and Pb isotope ratios obtained for igneous rocks from the shield are indicative of a mantle magma source. Although many of the dismembered ophiolites cannot be identified with inter-terrane sutures in their present location, the eastern margin of the Nabitah orogenic belt is a major tectonic break that coincides with a critical boundary between Nd and Pb isotopic provinces and is marked by a linear array of ophiolite fragments across the length of the shield. Other terrane boundaries have not been identified conclusively, both because coeval island arcs can not be distinguished readily on isotopic grounds and because many ophiolites are allochthonous. However, the calculated rates of crustal growth (measured as volume of magma, extracted from the mantle per unit time) between 900 and 600 Ma are similar to those calculated for Phanerozoic terranes from the Canadian Cordillera. Such high rates in the Afro-Arabian Shield suggest that island arc terranes have accreted along a continental margin now exposed in NE Africa, together with minor continental fragments. If crustal growth rates during this time were no greater than contemporary rates, *ca.* 4000 km of arc length are required, which is considerably less than that responsible for crustal growth in the SW Pacific.

1. Introduction

The recognition that some continental margins result from the accretion of lithospheric fragments, each with a distinctive stratigraphy and separated by tectonic contacts, was initially made in the North American Cordillera during the 1970s. Over 200 terranes have been recognized in the Cordillera which is now believed to be a collage of oceanic plateaus, intra-oceanic volcanic arcs and continental fragments, brought together over large distances (greater than 10^3 km) by oblique convergence (Oldow *et al.* 1989). This discovery had a profound influence on the interpretation of older tectonic margins, and led to a series of criteria being erected to identify allochthonous terranes by the recognition of lithological, structural, palaeomagnetic or faunal contrasts across terrane boundaries. Where faunal or palaeomagnetic data are available to aid reconstruction, or where inter-terrane boundaries have not been disturbed by subsequent tectonic activity, displaced terranes can often be readily recognized as in the North American Cordillera or the Tibetan Plateau. However, for Precambrian plate margins, none of these conditions commonly applies. Consequently, terrane recognition must be necessarily more speculative. This paper reviews the contribution that geochemical studies can make to the identification of terranes in general, and uses the Afro-Arabian Shield as a late Proterozoic example.

2. THE GEOCHEMISTRY OF TERRANE ACCRETION

The recognition of terranes rests either on the study of the tectonic boundaries separating the terranes or on the contrasting properties of the terranes themselves. In general, geochemical studies of igneous rocks can indicate the magma source, and by comparison with present-day magma geochemistry can identify the processes responsible for magmagenesis. For example, trace element studies of basic rocks can help to identify the tectonic setting in which they have formed (Pearce 1982). By using discriminant diagrams based on critical trace elements, ophiolite fragments can be identified as components of oceanic lithosphere and the nature of this lithosphere (for example MORB-type or supra-subduction zone) determined. Such ophiolites, if occurring within suture zones, will then provide constraints on the original tectonic setting of inter-terrane oceanic lithosphere which pre-dated docking. Information can also be obtained from granitic complexes that occur within terranes (Pearce et al. 1984). For example, their trace elements can indicate the presence of subduction-related magmatism which resulted from the pre-accretion closure of large ocean basins.

Isotopic dating of igneous rocks by using conventional U–Pb zircon or Rb-Sr whole rock techniques will determine periods of magmatic activity within each terrane. Where contrasting periods of magmatism occur within terranes, which then become juxtaposed through docking, simple age dating may constrain inter-terrane boundaries. Unfortunately, this is rarely if ever the case. Terranes commonly have overlapping periods of magmatism and idealized patterns may be subsequently complicated by post-collision magmagenesis. So, in general, the ages of magmatic provinces do not define terrane boundaries. Fortunately, magmatic rocks retain critical information on the age and nature of their source regions. In the case of Sm–Nd decay, model Nd ages can be calculated for igneous or sedimentary rocks which constrain when the rock or its crustal precursor was extracted from a mantle reservoir. If, for example, a magma formed directly from mantle melting the model Nd age will equal the emplacement age. In general, the model Nd age is somewhat older than the emplacement age reflecting the time of crust generation. The importance of Nd model ages rests on the fact that Sm and Nd, unlike Sr and Rb, are not readily fractionated by crustal processes but are fractionated between the crust and mantle. Thus Nd model ages record the time when the rock was extracted from the mantle (usually taken to be a depleted upper mantle) irrespective of subsequent periods of crustal reworking or sedimentary cycles. If several sources are involved, the Nd model age provides a weighted average of the ages of the contributing sources. Model Nd ages, therefore, provide direct information on the source of the magma, whether that is in the mantle or the continental crust.

Regional Nd-isotope studies map out isotopic variations in the basement, and sharp changes in such trends may reflect terrane margins where crust generation in adjacent terranes has occurred at different times. Pb isotopes also provide information on the source of the magma although Pb and Nd isotope studies offer distinct geochemical information because of the different behaviour of U:Pb, Th:U and Sm:Nd in different geological environments. In general, Sm:Nd is strongly fractionated between mantle and crust, whereas U:Pb is affected by intra-crustal fractionation and Th:U by sedimentary recycling. U:Pb ratios show much more variation in crustal rocks than do Sm:Nd, so that, whereas Pb isotopes studies are commonly successful in distinguishing between different crustal processes, periods of crustal growth are better identified through Nd isotopes.

Where areas of strongly contrasting periods of crust generation are juxtaposed by terrane tectonics, isotope geochemistry may reveal inter-terrane boundaries. For example, in southern California, approximately north–south isopleths of initial Sr ratios have been mapped in the Sierra Nevada batholith, where the rapid variation between 0.704 and 0.706 is correlated with Pb isotope and trace element variations within the batholith, and with a major fault zone (Kistler & Peterman 1973). This shift in initial isotopic ratios has been recognized as a suture between oceanic and continental basement terranes (Saleeby 1981).

For pre-Phanerozoic terranes, isotope geochemistry may provide the only available criteria distinguishing terranes, but caution is required if isotopic provinces are to be correlated with discrete terranes. Far less is known about the growth and tectonics of the Proterozoic of the western United States than of the Mesozoic of the North American Cordillera. However, isotopic basement trends in the western United States have been mapped for both Nd and Pb isotopic ratios. Nd model ages are found to conform to a general pattern (figure 1 a) that reflects a systematic decrease in model age away from the Archaean nucleus in the north (Bennett & DePaolo 1987). The data are consistent with crustal growth at a continental margin, active between 1.7 and 1.5 Ga. The lack of sharp boundaries in the isotopic characteristics of the basement and the lack of correlation between basement geochemistry and major faults or shear zones argue against their distribution being controlled by tectonic boundaries.

Early Proterozoic outcrop is not extensive. Magmatic ages in the region range from 1700 to 60 Ma and their distribution shows no simple trends (figure 1 a). The importance of model Nd

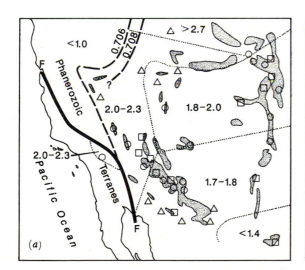 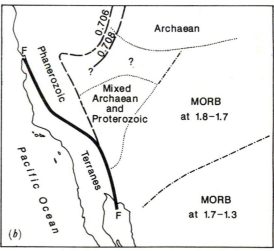

FIGURE 1. Isotopic provinces from the Proterozoic basement of the western United States. (a) Nd isotope provinces (Bennett & De Paolo 1987) with characteristic model Nd ages in Ga. Dotted lines delimit model Nd age provinces (initial ^{87}Sr:^{86}Sr isopleths at 0.706, 0.708 taken from Kistler & Peterman (1973)). Stippled areas, early Proterozoic outcrop. Symbols indicate sample locations with the following crystallization ages: ○, 1.7 Ga; □, 1.4 Ga; △, 0.06 Ga. (b) Pb isotope provinces (dotted line from Wooden et al. (1986); dot-dash from Oldow et al. (1989)).

age trends is that they reflect the character of the basement, and effectively 'see-through' subsequent intra-crustal processes such as post-accretion magmatism.

The boundaries between Archaean and Proterozoic Pb isotope provinces are consistent with Nd isotope mapping (figure 1*b*), but within the region of Proterozoic basement, the distribution of Pb provinces is more equivocal. These were initially explained as three or more terranes of different characteristics that had accreted during a period of rapid crustal growth 1.8–1.6 Ga ago (Wooden *et al.* 1986). A terrane model is favoured by Condie (1986) partly on the basis of irregular age distributions for felsic eruptions between 1.7 Ga and 2.3 Ga. The redrawing of some of the boundaries between Pb isotopic provinces (Oldow *et al.* 1989) suggests a systematic shift of less radiogenic lead outwards from the Archaean nucleus (figure 1*b*). As for the Nd data, this pattern is consistent with a model of crustal growth at a continental margin rather than a random accretion of allochthonous terranes.

Terrane accretion is a mechanism of lateral growth for continents. Since continental lithosphere is not vertically homogenous, it cannot be assumed either that variations in initial isotopic ratios of crustal melts indicate a terrane boundary (they may reflect a lateral change in depth from which magma is derived) or conversely that lack of variation rules out a terrane boundary. For example, in one study across a known Archaean–Proterozoic boundary in Wyoming, Proterozoic Nd model ages were obtained from granitoids on both sides of the boundary (Geist *et al.* 1988). The implication is that the Archaean continental crust had been underplated near the suture during Proterozoic magmagenesis. What is important in relating isotopic provinces to terrane accretion is the pattern of their distribution and their correlation with major tectonic features.

Nd model ages are also of value for sedimentary rocks. In fine-grained clastic sediments, Nd model ages provide an average age of their source regions: that is a weighted average of the times at which contributing sources to the detritus were extracted from the mantle. Consequently, where terranes of contrasting basement ages are brought together, a marked shift in model ages should be observed in sediments deposited after the docking of the terranes. This has been observed in northern Britain where sediments deposited south of the Iapetus suture, which represents closure of a major ocean, show an increase in Nd model ages for deposition younger than about 500 Ma (Davies *et al.* 1985; Miller & O'Nions 1984). This is persuasive geochemical support for a tectonic model that argues for the arrival of an older terrane as a consequence to ocean closure. Unfortunately, such an elegant technique requires well-constrained deposition ages which are seldom available in Precambrian shields.

3. THE AFRO-ARABIAN SHIELD

The late Proterozoic crystalline basement of the Afro-Arabian shield is now separated into Arabian and African segments by the late Tertiary opening of the Red Sea, and is onlapped to the East and West by Phanerozoic sediments. The exposed geology of the Shield (figure 2) is dominated by granitic intrusions (greater than 66%) and calc-alkaline volcanics with relatively minor volcanoclastic sediments, all of Pan-African (900–600 Ma) age. Mafic-ultramafic complexes occurring therein have been identified as ophiolitic sequences (Bakor *et al.* 1976; Al Shanti & Mitchell 1976; Gass 1977). Moreover, a growing body of geochemical data identifies the great majority of magmatic rocks as originating in an arc (supra-subduction zone) environment and consequently the subparallel ophiolite zones have been interpreted as

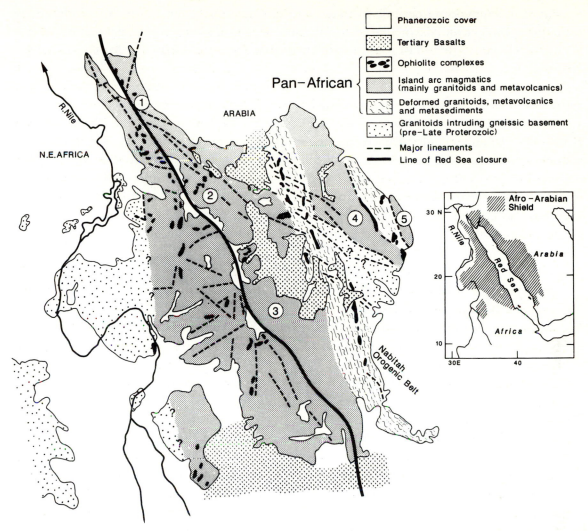

FIGURE 2. Sketch geological map of Afro-Arabian shield. Insert shows present distribution of basement outcrops. Arabian 'terranes' after Stoeser & Camp (1985); 1, Midian; 2, Hijaz; 3, Asir; 4, Afif; 5, Ar Rayn.

marking inter-arc sutures. In this model, the arcs accreted to the African craton by the end of the Pan-African to form the Afro-Arabian Shield (Greenwood *et al.* 1976; Gass 1977).

Analogies have been drawn between the terrane accretion of the North American Cordillera and the tectonics of the Afro-Arabian Shield (Kroner 1985). Ophiolite-decorated sutures are identified as the boundaries between geologically discrete terranes. Indeed, a hierarchy of terranes has evolved in the literature for the Arabian segment of the shield; Johnson & Vranas (1984) recognized 10 terranes, which were bound by major faults and ophiolite 'lineations' referred to as sutures. Each terrane was described as a coherent sedimentary and volcanic environment, and igneous suites within each terrane shared petrologic and chemical affinities. Stoeser & Camp (1985) provided a simpler model of five terranes in Arabia separated by four sutures. This terrane hierarchy was further subdivided by Stoeser & Stacey (1988). However, all subdivisions of the shield recognize the importance of the Nabitah orogenic belt; a 200 km wide zone of syn-orogenic granite plutons separated by regions of amphibolite grade metasediments and metavolcanics (Droop & Al-Filali 1989). This separates western terranes, which are unequivocally oceanic in character, from eastern terranes, which include at least some pre-Pan-African continental material. The Stoeser & Camp model defined a period of island arc evolution (950–715 Ma) followed by terrane accretion (715–630 Ma) and finally by within plate magmatism (630–550 Ma).

Unfortunately, with crust of early Phanerozoic–late Precambrian age the matching of faunal provinces to identify terranes is not possible. Moreover, palaeomagnetic reconstructions, used so successfully in identifying allochthonous terranes in North America, have not been successful in Arabia. The only detailed palaeomagnetic study of the area (Kellogg & Beckman 1984) was undertaken in the southeastern Arabian Shield; this concluded that pre-600 Ma magnetic signatures were destroyed by the widespread Pan-African granite plutonism. By this time, the apparent polar wandering path of Arabia is indistinguishable from that of Africa. There is, therefore, no palaeomagnetic information on the relative positions of different terranes in the shield before 600 Ma and their recognition rests almost entirely on geochemical criteria.

(a) Ophiolites and inter-terrane sutures

The ophiolite sequences of the Afro-Arabian shield include ophiolitic melanges, serpentinite thrust melanges and fault-bounded inliers. To interpret the outcrop distribution of ophiolite fragments as terrane boundaries or sutures, it must be assumed that the ophiolites have not been displaced significantly from the suture during obduction or post-obduction collisional processes.

There are several lines of evidence that suggest that the present-day distribution of many of the ophiolite fragments from the Afro-Arabian Shield do not mark ancient sutures. Afro-Arabian ophiolites form detached masses with tectonic contacts and there is no evidence for primary obduction such as the metamorphic sole of ocean floor basalts to the Tethyan Oman ophiolite (Lippard et al. 1986). Detailed structural studies of ophiolites from the Eastern Desert of Egypt recognize the ophiolite fragments as allochthonous ophiolitic melanges, possibly displaced from their suture zone by several hundred kilometres (Ries et al. 1983). Moreover, in the Arabian Shield the ophiolite distribution has been rearranged by the Nadj fault system; a left-lateral shear zone with 200–300 km offsets which evolved between 630–550 Ma (Davies 1984) and clearly post-dated all obduction.

Perhaps one criterion for associating ophiolites with sutures is their virtual continuity over many hundreds of kilometres as in the Tethyan suture zone of southern Tibet. The only linear zone of ophiolitic fragments in the Afro-Arabian shield that may be traced over such distances is in the Nabitah orogenic belt (figure 2). Moreover the Pan-African ophiolites are smaller and more widely spaced than their Tethyan counterparts. This may simply be an erosional effect for virtually all the Semail ophiolite of Oman would disappear if the area were eroded to a depth of 5 km (Lippard et al. 1986). Maximum depths of erosion in the Afro-Arabian Shield as measured in the Nabitah Orogenic belt are about 10 km (Droop & Al Filali 1989) although much of the southern portion of the shield has only been subjected to about 5 km of erosion (Gass et al. 1990).

As well as indicating the site of former oceans, ophiolites identify the composition of that oceanic lithosphere and, thereby, its tectonic setting. Trace elements from the basic eruptive components together with their field associations suggest that most of the ophiolites in both northeast Africa and Arabia were generated in a supra-subduction zone tectonic setting as, for example, in Jebel al Wask (Bakor et al. 1976), Jebel Thurwah (Nassief et al. 1984), and the

Halieb ophiolite NE Sudan (Price 1984). Ophiolites of the Al Amar zone in eastern Arabia have boninitic characteristics, again indicating a supra-subduction zone setting (Al Shanti & Gass 1983). Furthermore, precise single zircon studies of ophiolites from the Egypt–Sudan border dated at 740–750 Ma are generally younger than the oldest magmatic arc sequences (850–870 Ma) and may therefore derive from intra-arc basins (Kroner & Todt 1989). It appears, therefore, that these ophiolites probably do not mark the closures of wide ocean basins but those of several small marginal basins such as are found presently in the western Pacific.

(b) Crystallization ages and terranes

Of the many Rb–Sr whole rock isochrons published from the Afro-Arabian Shield, few reliable isochrons have been recorded outside the range 850–550 Ma. Spatially related age trends are not apparent within the shield although a general trend of younging of arc-related magmatism from north to south in the western terranes has been suggested (Stoeser & Camp 1985). Variations of initial (^{87}Sr:^{86}Sr) ratios are virtually restricted to between 0.707 and 0.702 throughout the shield and any spatial trend within this range is masked by analytical uncertainties. Stoeser & Camp (1985) record a slightly higher range (0.707–0.7035) east of the Nabitah orogenic belt compared with values of 0.7035–0.702 to the west. However, a detailed study of volcanics from the south eastern shield provided no initial ratios in excess of 0.7036 (Darbyshire et al. 1983). Therefore, neither Rb–Sr crystallization ages nor initial Sr isotope ratios provide unequivocal evidence for discrete terranes within the Afro-Arabian shield.

Numerous zircon studies also reveal a general age span of 870–550 Ma within the shield (Stoeser & Stacey 1988 and references therein). There are, however, some interesting examples of pre-Pan-African ages, which have been reviewed by Harris et al. (1984) and Kroner et al. (1988). In the south of the eastern terranes of Arabia, three upper intercept zircon ages from magmatic rocks lie in the range 1800–1600 Ma indicating the presence of some pre-Pan-African basement. In Africa, early Proterozoic zircon ages are found in sedimentary rocks from the Eastern Desert of Egypt and to the southeast of Lake Nasser. There is no doubt that a pre-Pan-African source region is required for some sediments, which must lie west of the Nile. An Archaean craton of unknown size certainly exists in the Uweinat inlier on the Egyptian–Libyan border where Rb–Sr isochrons of 2600 Ma have been obtained (Klerkx & Deutsch 1977). Much further south, ion microprobe ages of 1000–2700 Ma from zircons in the Sabaloka basement north of Khartoum (Kroner et al. 1987) confirm the presence of a mid-Proterozoic or Archaean source region in western Sudan.

Crystallization ages indicate the presence of pre-900 Ma ages in the southeastern Arabia and west of the Nile but do not define discrete age provinces within the Afro-Arabian shield. There remain two isotopic techniques that provide information not on the periods of magmatism but on the age and geochemistry of the underlying basement.

(c) Pb–Pb isotopic data

Pb isotopic data may be plotted on two evolution diagrams (figure 3). For initial ratios, the upper plot (^{208}Pb:^{204}Pb against ^{206}Pb:^{204}Pb) reflects variations in Th:Pb and U:Pb and hence the Th:U ratio in the source. The Pb isotope evolution through time of average upper mantle, as modelled by Zartman & Doe (1981), is also plotted. The growth curve for average crust is not indicated since crustal Th:U ratios are extremely variable but, in general, sedimentary recycling will result in elevated Th:U, resulting in fields above the mantle growth curve for

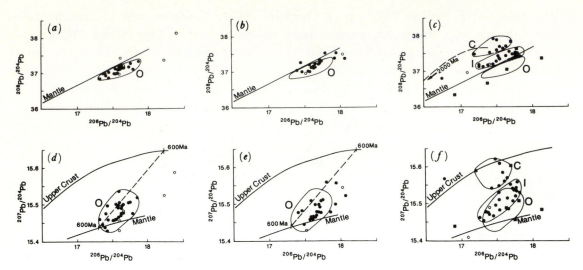

FIGURE 3. Pb isotopic characteristics of the Afro-Arabian shield with modelled upper mantle and upper crust evolution from Zartman & Doe (1981). Closed symbols indicate data from sediments and intermediate-acid magmatic rocks. Open symbols, ophiolites. (a), (d) The western terranes of Arabia. Dashes indicate mixing line between upper crust and mantle at 600 Ma. (b), (e) The Nabitah orogenic belt. (c), (f) Eastern terranes of Arabia. Squares indicate samples from NE Africa. Dashed line indicates evolution line for crust (U:Th = 4.0) extracted from mantle at 2000 Ma. Empirical fields indicated by O (oceanic Pb), I (intermediate Pb) and C (continental Pb). (Data from Stacey et al. 1980; Stacey & Stoeser 1983; Gillespie & Dixon 1983; Pallister et al. 1988; Stoeser & Stacey 1988; Ellam et al. 1990.)

mature continental crust (Ellam et al. 1990). For the lower plot ($^{207}Pb:^{204}Pb$ against $^{206}Pb:^{204}Pb$), mantle and crustal reservoirs are better constrained since fractionation of crust from upper mantle results in at least a slight increase in U:Pb, and modelled evolution curves for both average upper mantle and average upper crust are given from Zartman & Doe (1981). Also shown is a dashed line linking the isotopic ratios of the model crust and mantle at 600 Ma. Interpretation of Pb isotope ratios is often equivocal since there is considerable variation in both mantle and crustal sources. The upper mantle is not only heterogeneous but, at island arcs, $^{207}Pb:^{204}Pb$ ratios in particular can be elevated by small components of subducted sediment.

Data from the Afro-Arabian Shield are from galena and feldspar separates which record the initial Pb isotope ratio at the time of crystallization. In the western Arabian terranes (figure 3a) samples cluster just below the mantle growth curve suggesting that the crust has primitive or oceanic isotopic ratios (field O). For the Nabitah orogenic belt (figure 3b) samples generally lie in the same field, but for the eastern Arabian terranes samples have a higher initial $^{208}Pb:^{204}Pb$ suggesting elevated Th:Pb and Th:U in their source (figure 3c) and therefore that mature continental material has contributed to their source regions. These data have been divided empirically into field I (intermediate) and field C (continental) and the spatial distribution of samples with O, I and C characteristics are plotted on figure 4a. In figure 3c a two-stage model is illustrated where crust is extracted from the upper mantle at 2000 Ma with an increased Th:U ratio of 4, compared with a typical mantle value of 3.5 (2000 Ma is chosen because it represents a typical model Nd age for samples in field C). Field I could represent either mixed sources of O and C type or a discrete terrane of pre-Pan-African age of less than 2000 Ma. In either event, the data identify an older continental fragment in the south of the eastern Arabian terranes and, although Pb data from Africa are sparse, there is some evidence for a pre-Pan-African continent in southwestern Egypt (figure 4a).

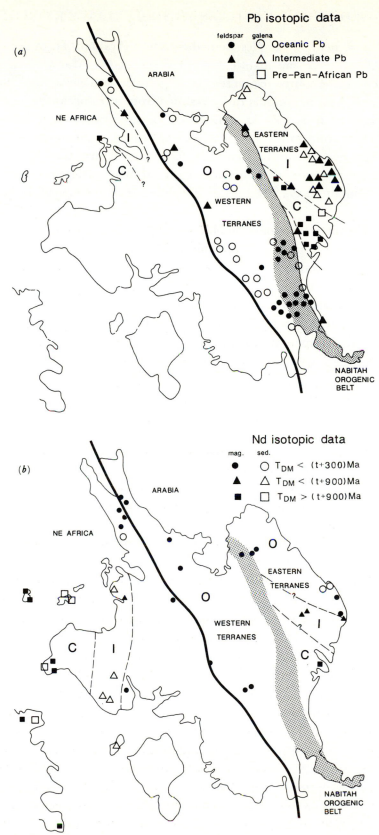

FIGURE 4. Isotopic data for Afro-Arabian shield. (*a*) Distribution of Pb data from galena (open symbols) and feldspar (closed symbols). ○, Oceanic (MORB) Pb at 600 Ma; △, intermediate Pb; □, continental Pb (as defined in figure 3). (*b*) Distribution of Nd initial ratios from magmatic rocks (closed symbols) and sediments (open symbols). ○, Oceanic Nd; △, intermediate Nd; □, continental Nd (as defined in figure 5).

In figure 3d and e, field O samples from the eastern Arabian Terranes and the Nabitah orogenic belt scatter above the mantle growth curve at 600 Ma, with some evidence for subducted sediment in the source region seen from samples with elevated ^{207}Pb:^{204}Pb ratios. It is proposed that such oceanic isotopic ratios resulted from intra-oceanic arcs between 900 and 600 Ma ago. However, in eastern Arabia, there is a significant shift to more crustal values (figure 3f), and these confirm a Lower Proterozoic crustal fragment in the south-east of the Arabian Shield. However, it can be seen from comparing figures 2 and 4a, that, if the I and C isotopic provinces do mark discrete terranes, their boundaries only coincide with those determined from ophiolite-decorated sutures along the eastern margin of the Nabitah Orogenic Belt.

In general, Pb isotopic ratios from ophiolite samples lie in the fields defined by samples from the adjacent terranes. In other words, ophiolites from the western Arabian terranes lie below the mantle growth curve in the ^{208}Pb:^{204}Pb against ^{206}Pb:^{204}Pb plot (figure 3a, b) but from the western Arabian terranes lie above the mantle growth curve (figure 3c) which is consistent with the generation of ophiolites in a supra-subduction environment. The Pb isotope ratios of ophiolites from the eastern Arabian terranes probably have been affected by subduction of some pre-Pan-African sedimentary material.

(d) Nd *isotopic data*

Model Nd ages (T_{DM}) from the Afro-Arabian shield are plotted in figure 5 against their emplacement age (t). t represents the crystallization age for igneous rocks and deposition age for sediments. The majority of samples from Arabia have emplacement ages approximately equal to their model Nd ages ($T_{DM} - t$) < 300 Ma. This suggests a short time interval between extraction of the melt from the mantle and its emplacement. However, for samples from eastern Arabia and Africa, there is a wider interval between model Nd age and emplacement age. Unfortunately, Pb and Nd isotope data are rarely available on the same samples, but Nd data can also be divided into three fields, O, I and C with increasing ($T_{DM} - t$) (figure 5) and their spatial distribution is plotted in figure 4b. They exhibit an E–W variation in Nd isotopic characteristics across NE Africa. On the Red Sea coast, sedimentary and plutonic samples essentially have 'oceanic' characteristics with Nd model ages less than 1100 Ma. To the west of this belt, both plutonic and sedimentary rocks have increased model Nd ages (1700–1000 Ma) and fall in field I defined by ($T_{DM} - t$) lying within the range 300–900 Ma. The eastern boundary of samples with these characteristics, roughly coincides with the river Nile. These can be interpreted either as representing a discrete continental terrane of Proterozoic age, or a continental margin to which both juvenile and Archaean continental sources contributed. A detailed study of interlayered metavolcanics and metasediments from southern Egypt (Wust *et al.* 1987) reveals much older Nd model ages for the metasediments (1700 Ma) than for the metavolcanics (*ca.* 800 Ma) suggesting that the sediments are derived from an ancient distal source, probably an Archaean craton to the west.

The presence of a North African craton is well-documented by samples from western Egypt and central Sudan which define a third group of continental character (field C, figure 5) in which model Nd ages range from 2500 to 1600 Ma. Further west Archaean model Nd ages of over 3000 Ma are obtained (Harris *et al.* 1984). The size of this craton is unknown, although it certainly extends westwards for several hundred kilometres and north–south for at least the length of the Afro-Arabian Shield (1200 km). What is important about the pattern of Nd

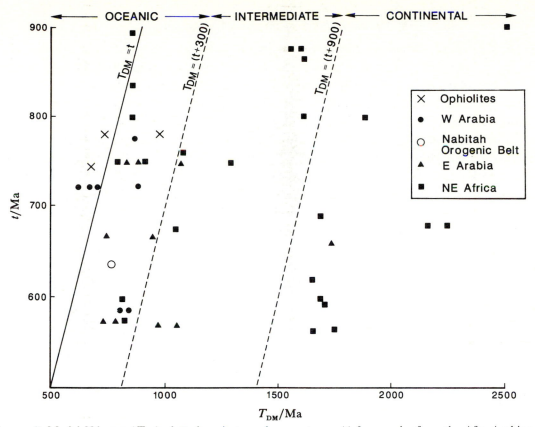

FIGURE 5. Model Nd ages (T_{DM}) plotted against emplacement ages (t) for samples from the Afro-Arabian shield. (Data from Bokhari & Kramers 1981; Duyverman *et al.* 1982; Claessons *et al.* 1984; Stacey & Hedge 1984; Harris *et al.* 1984; Reischmann *et al.* 1986; Kroner *et al.* 1987; Wurst *et al.* 1987; Schandelmeier *et al.* 1988.)

results in Northeast Africa is that it records a transition between cratonic and oceanic characteristics which probably reflects a continental margin, but the spatial distribution of Nd model ages has no relation either to major lineaments or to the present-day distribution of ophiolite fragments (figure 2).

For the Arabian Shield, there are fewer Nd data but the distribution of available Nd data essentially mimics the pattern better defined by the Pb isotope data. Samples with oceanic Nd signatures characterize the western terranes of Arabia, including the Nabitah orogenic belt. East of the Nabitah orogenic belt, intermediate model ages occur in the centre of the exposed shield but more oceanic characteristics are located further east and north. In the south, a single early Proterozoic model age has been published although more samples with older model Nd ages have been obtained both from this region and from North Yemen (J. S. Stacey, personal communication).

In general, there is not a strong correlation between the distribution of Pb and/or Nd isotopic provinces and the distribution of ophiolite fragments. The only unambiguous example in the Afro-Arabian shield of ophiolite outcrops coinciding with the boundary between age provinces lies along the eastern margin of the Nabitah orogenic belt. Elsewhere in the shield, terranes may exist but if, as seems likely from major and trace element geochemistry, the shield results from accreted island arcs, their similar periods of formation prevents their being resolved isotopically.

(e) Crustal evolution of the Afro-Arabian Shield

The Afro-Arabian Shield is continental crust of normal thickness with geochemical characteristics indicative of crustal growth by late Proterozoic subduction processes. It is possible to calculate the rate at which continental crust was generated in this area and to compare it with contemporary rates at active subduction zones (Reymer & Schubert 1982, 1984; Duyverman *et al.* 1982; Dixon & Golombek 1988). However, such calculations are poorly constrained because of large uncertainties in both the surface area of the Pan-African crust and in the length of arcs active over the 300 Ma period. The area of Pan-African crust, if restricted to known outcrop of Pan-African age, is approximately 1.2×10^6 km², equivalent to a volume of 4.8×10^7 km³ (assuming an average crustal thickness of 40 km) which implies an average growth rate of 0.16 km³ a⁻¹ over 300 Ma. Present-day global rates are 1.0 km³ a⁻¹, or 20–40 km³ Ma⁻¹ per kilometre of arc length (Reymer & Schubert 1984). If we assume similar growth rates per kilometre of arc operated in the late Proterozoic (geothermal gradients were only about 10 % greater 900 Ma ago compared with current values) then 4000 km of arc are required to account for the exposed Afro-Arabian island arc terranes during the Pan-African event.

Over 1200 km of arc can be inferred by adding the lengths of ophiolite sutures in Arabia alone (Stoeser & Camp 1984) and a value of 3600 km has been obtained from identifications of liner granitic zones throughout the Afro-Arabian shield (Gass 1982). The 4000 km of arc implied by crustal growth rates, based on isotope systematics, is a reasonable estimate and is broadly consistent with arc lengths estimated from field studies.

The total area of Pan-African crust is probably much greater than is presently exposed. The presence of calc-alkaline granitoids of Pan-African age with low initial (^{87}Sr:^{86}Sr) ratios in the Oman suggest much of the Phanerozoic sediment of Arabia may be underlain by Pan-African basement (Gass *et al.* 1990). In NE Africa, the Pan-African crust extends at least as far south as Ethiopia and possibly into northern Somalia (Kroner *et al.* 1989). Overall, the total area of Pan-African crust in the Afro-Arabian shield may be as large as 3.5×10^6 km² although the proportion of older terranes within this extensive area remains unknown. This larger area implies a crustal growth rate of 0.47 km³ a⁻¹, which is almost 50 % of that generated globally at the present time. The total length of arc required for this increased volume is about 12 000 km considerably less than the 23 000 km active in the SW Pacific over the past 200 Ma (Reymer & Schubert 1986).

High crustal growth rates in a relatively small area are powerful evidence for a terrane-based tectonic model. A terrane collage as exposed in the North American Cordillera, or as presently being accreted in Southeast Asia will provide high crustal production rates because only in such a setting will arcs formed across wide oceans become juxtaposed. In contrast, crustal growth by magmatic accretion at an active continental margin is restricted to an arc length equivalent to the length of the margin. Moreover a significant proportion of magmatism will include a recycled component (*ca.* 20 % according to Bennett & DePaolo (1987)) thus reducing growth rates calculated from isotopic criteria. A recent Nd study of the Canadian Cordillera identified five large terranes of oceanic affinity which indicate a crustal growth rate of 0.13 km³ a⁻¹ (Samson *et al.* 1989), similar to the minimum calculated rate for the Afro-Arabian Shield. If the distribution of isotopic provinces has not provided conclusive evidence for terrane tectonics in the Afro-Arabian Shield, the high crustal growth rate is difficult to explain by any other process.

4. CONCLUSIONS

Geochemical techniques can provide important constraints on terrane accretion, particularly where palaeomagnetic and faunal evidence are not available.

Trace element studies of both ophiolite fragments and more evolved magmatic rocks can identify the tectonic régime in which oceanic lithosphere and continental crust has formed. Isotopic analysis of sedimentary rocks can identify the geochemical characteristics of sedimentary source regions and thereby the timing of docking between terranes provided that uplifted regions in adjacent terranes are isotopically distinct.

In Precambrian shields it is commonly not possible to determine the age of sedimentation with sufficient accuracy for this technique to be of value, but here isotopic provinces defined by Pb and Nd isotope ratios can provide information on the chemical characteristics of the basement. Where sharp isotopic variations coincide with structural discontinuities then a terrane boundary may be established, as across the eastern margin of the Nabitah belt in the Arabian shield. Such studies will not determine distances over which adjacent terranes have been displaced but that they have evolved in distinct tectonic environments.

Terrane accretion can lead, locally, to unusually high crustal growth rates if the collage incorporates a high proportion of island arc fragments. In the Afro-Arabian shield, high crustal generation rates are broadly similar to those calculated for some juvenile terranes from the North American Cordillera. Along such continental margins, a high proportion of the global crustal growth rate may appear to have occurred across distances of a few hundred kilometres although, at the time of crustal generation, the island arcs responsible for crustal growth may have been widely separated.

REFERENCES

Al-Shanti, A. M. S. & Mitchell, A. H. G. 1976 Late Precambrian subduction and collision in the Al Amar–Idsas region, Arabian shield, Kingdom of Saudi Arabia. *Tectonophysics* **31**, 41–47.

Al-Shanti, A. M. S. & Gass, I. G. 1983 The Upper Proterozoic ophiolite melange zones of the easternmost Arabian shield. *J. geol. Soc. Lond.* **140**, 867–876.

Bakor, A. R., Gass, I. G. & Neary, C. R. 1976 Jabal Al-Wasq, northwest Saudi Arabia: an Eo-cambrian back-arc ophiolite. *Earth planet. Sci. Lett.* **30**, 1–9.

Bennett, V. C. & DePaolo, D. J. 1987 Proterozoic crustal history of the western United States as determined by Nd isotope mapping. *Geol. Soc. Am. Bull.* **99**, 674–685.

Bokhari, F. Y. & Kramers, J. D. 1981 Island arc character and Late Precambrian age of volcanics at Wadi Shwas, Hijaz, Saudi Arabia: geochemical and Sr and Nd isotopic evidence. *Earth planet. Sci. Lett.* **54**, 409–422.

Claesson, S., Pallister, J. S. & Tatsumoto, M. 1984 Sm–Nd data on two late Proterozoic ophiolites of Saudi Arabia and implications for crustal and mantle evolution. *Contr. Mineral. Petrol.* **85**, 244–252.

Condie, K. C. 1986 Geochemistry and tectonic setting of Early Proterozoic supracrustal rocks in the SW United States. *J. Geol.* **94**, 845–864.

Darbyshire, D. P. F., Jackson, N. J., Ramsay, C. R. & Roobol, M. J. 1983 Rb-Sr isotope study of latest Proterozoic volcano-sedimentary belts in the Central Arabian Shield. *J. geol. Soc. Lond.* **140**, 203–213.

Davies, F. B. 1984 Strain analysis of wrench faults and collision tectonics of the Arabian-Nubian shield. *J. Geol.* **82**, 37–53.

Davies, G., Gledhill, A. & Hawkesworth, C. J. 1985 Upper crustal recycling in southern Britain: evidence from Nd and Sr isotopes. *Earth planet. Sci. Lett.* **75**, 1–12.

Dixon, T. H. & Golombek, M. P. 1988 Late Precambrian crustal accretion rates in northeast Africa and Arabia. *Geology* **16**, 991–994.

Droop, G. J. R. & Al-Filali, I. Y. 1989 Magmatism, deformation and high-*T*, low-*P* regional metamorphism in the Nabitah Mobile Belt, southern Arabian Shield. In *Geol. Soc. Lond., Spec. Rep.* **43**, 469–48, *Evolution of metamorphic belts* (ed. J. F. Daly, R. A. Cliff & B. W. D. Yardley).

Duyverman, H. J., Harris, N. B. W. & Hawkesworth, C. J. 1982 Crustal accretion in the Pan-African: Nd and Sr isotope evidence from the Arabian Shield. *Earth Planet. Sci. Lett.* **59**, 315–326.

Ellam, R. M., Hawkesworth, C. J. & McDermott, F. 1990 Pb isotope data from late Proterozoic subduction-related rocks: implications for crust-mantle evolution. *Chem. Geol.* (In the press.)

Gass, I. G. 1977 The evolution of the Pan-African crystalline basement in NE Africa and Arabia. *J. geol. Soc. Lond.* **134**, 129–138.

Gass, I. G. 1982 Upper Proterozoic (Pan-African) calc-alkaline magmatism in north-eastern Africa and Arabia. In *Andesites* (ed. R. S. Thorpe), pp. 571–609. Wiley.

Gass, I. G., Ries, A. C., Shackleton, R. M. & Smewing, J. D. 1990 Tectonics, geochronology and geochemistry of the Precambrian rocks of Oman. In *Geol. Soc. Lond., Spec. Rep. Geology and tectonics of the Oman region* (ed. A. H. F. Robertson & A. C. Ries) **49**, pp. 585–599.

Geist, D. J., Frost, C. D., Kolker, A. & Frost, B. R. 1988 A geochemical study of magmatism across a major terrane boundary: Sr and Nd isotopes in Proterozoic granitoids of the southern Laramie Range, Wyoming. *J. Geol.* **97**, 331–342.

Gillespie, J. G. & Dixon, T. H. 1983 Lead isotope systematics of some igneous rocks from the Egyptian Shield. *Precambrian Res.* **20**, 63–77.

Greenwood, W. R., Anderson, R. E., Fleck, R. J. & Schmidt, D. L. 1976 Late Proterozoic cratonisation in southwestern Arabia. *Phil. Trans. R. Soc. Lond.* A **280**, 517–527.

Harris, N. B. W., Hawkesworth, C. J. & Ries, A. C. 1984 Crustal evolution in north-east Africa from model Nd ages. *Nature, Lond.* **309**, 773–776.

Johnson, P. R. & Vranas, G. J. 1984 The origin and development of Late Proterozoic rocks of the Arabian Shield. *Directorate General Min. Res., Jeddah Open File Rep.* **04-32**.

Kellogg, K. S. & Beckman, E. J. 1984 Palaeomagnetic investigations of Upper Proterozoic rocks in the Eastern Arabian Shield, Kingdom of Saudi Arabia. *Faculty Earth Sci. Bull. King Abdulaziz University Jeddah* **6**, 484–500.

Kistler, R. W. & Peterman, Z. E. 1973 Variations in Sr, Rb, K, Na, and initial 87Sr/86Sr in Mesozoic granitic rocks and intruded wall rocks in central California. *Geol. Soc. Am. Bull.* **84**, 3489–3512.

Klerkx, J. & Deutsch, S. 1977 Resultats preliminaries obtenues par la methode Rb/Sr sur l'age des formations precambriennes de la region d'Uweinat (Libye). *Mus. R. Afr. centr. Tervuren Dept. Geol. Min. Rapp. ann. 1976*, pp. 83–94.

Kroner, A. 1985 Ophiolites and the evolution of tectonic boundaries in the Late Proterozoic Arabian-Nubian shield of Northeast Africa and Arabia. *Precambrian Res.* **27**, 277–300.

Kroner, A., Stern, R. J., Dawoud, A. S., Compston, W. & Reischmann, T. 1987 The Pan-African continental margin in northeastern Africa: evidence from a geochronological study of granulites at Sabaloka, Sudan. *Earth planet. Sci. Lett.* **85**, 91–104.

Kroner, A., Reischmann, T., Wust, H. J. & Rashwan, A. A. 1988 Is there any pre-Pan-African (> 950 Ma) basement in the Eastern Desert of Egypt? In *The Pan African Belt of NE Africa and adjacent areas* (ed. S. El Gaby & R. Greiling), pp. 95–120. Braunschweig: Vieweg.

Kroner, A. Egal, M., Egal, Y. Sassi, F. & Taklag, M. 1989 Extensions of the Arabian-Nubian Shield into southern Israel, Ethiopia and northern Somalia. *Terra. Abstr.* **1**, 363.

Kroner, A. & Todt, W. 1989 Dating of Late Proterozoic ophiolites in Northeast Africa by single zircon evaporation. *Terra Abstr.* **1**, 363.

Lippard, S. J., Shelton, A. W. & Gass, I. G. 1986 The ophiolite of Northern Oman. *Geol. Soc. Lond. Mem.* **11**. (178 pages.)

Miller, R. G. & O'Nions, R. K. 1984 The provenance and crustal residence ages of British sediments in relation to palaeogeographic reconstructions. *Earth planet. Sci. Lett.* **68**, 459–470.

Nassief, M. O., Macdonald, R. & Gass, I. G. 1984 The Jebel Thurwah Upper Proterozoic ophiolite complex, western Saudi Arabia. *J. geol. Soc. Lond.* **141**, 537–546.

Oldow, J. S., Bally, A. W., Ave Lallemant, H. G. & Leeman, W. P. 1989 Phanerozoic evolution of the North American Cordillera; United States and Canada. In *The geology of North America – an overview*, vol. A. *Geol. Soc. Am. Spec. Publ.*, pp. 139–232.

Pallister, J. S., Stacey, J. S., Fischer, L. B. & Premo, W. R. 1988 Precambrian ophiolites of Arabia: geologic settings, U–Pb geochronology, Pb-isotope characteristics, and implications for continental accretion. *Precambrian Res.* **38**, 1–54.

Pearce, J. A. 1982 Trace element characteristics of lavas from destructive plate boundaries. In *Andesites* (ed. R. S. Thorpe), pp. 525–548. Wiley.

Pearce, J. A., Harris, N. B. W. & Tindle, A. G. 1984 Trace element discrimination diagrams for the tectonic interpretation of granitic rocks. *J. Petrol.* **25**, 956–983.

Price, R. C. 1984 Later Precambrian mafic-ultramafic complexes in North-East Africa. Ph.D. thesis, Open University, Milton Keynes, U.K.

Reischmann, T., Kroner, A. & Hofmann, A. W. 1986 Nd isotope characteristics of Late Proterozoic rocks from the Red Sea Hills, Sudan, and their significance for the evolution of the Arabian-Nubian Shield (abstr). *Terra Cognita* **6**, 234.

Reymer, A. & Schubert, G. 1984 Phanerozoic addition rates to the continental crust and crustal growth. *Tectonics* **3**, 63–77.

Reymer, A. & Schubert, G. 1986 Rapid growth of some major segments of continental crust. *Geology* **14**, 299–302.

Ries, A. C., Shackleton, R. M., Graham, R. H. & Fitches, W. R. 1983 Pan-African structures, ophiolites and melange in the Eastern Desert of Egypt: a traverse at 26° N. *J. geol. Soc. Lond.* **140**, 75–95.

Saleeby, J. 1981 Ocean floor accretion and volcanoplutonic arc accretion of the Mesozoic Sierra Nevada. In *The geotectonic development of California, Rubey vol. 1* (ed. W. G. Ernst), pp. 132–181. Englewood Cliffs, New Jersey: Prentice-Hall.

Samson, S. D., McClelland, W. C., Patchett, P. J., Gehrels, G. E. & Anderson, R. G. 1989 Evidence from Nd isotopes for mantle contributions to Phanerozoic crustal genesis in the Canadian Cordillera. *Nature, Lond.* **337**, 705–709.

Schandelmeier, H., Darbyshire, D. P. F., Harms, U. & Richter, A. 1988 The East Saharan Craton: evidence for pre-Pan African crust in NE Africa west of the Nile. In *The Pan African Belt of NE Africa and adjacent areas* (ed. S. El Gaby & R. Greiling), pp. 69–94. Braunschweig: Vieweg.

Stacey, J. S., Doe, B. R., Roberts, R. J., Delevaux, M. H. & Gramlich, J. W. 1980 A lead isotope study of mineralisation in the Saudi Arabian Shield. *Contr. Mineral. Petrol.* **74**, 175–188.

Stacey, J. S. & Stoeser, D. B. 1983 Distribution of oceanic and continental leads in the Arabian-Nubian Shield. *Contr. Mineral. Petrol.* **84**, 91–105.

Stacey, J. S. & Hedge, C. E. 1984 Geochronologic and isotopic evidence for early Proterozoic crust in the eastern Arabian Shield. *Geology* **12**, 310–313.

Stoeser, D. B. & Camp, V. E. 1985 Pan-African microplate accretion of the Arabian Shield. *Geol. Soc. Am. Bull.* **96**, 817–826.

Stoeser, D. B. & Stacey, J. S. 1988 Evolution, U–Pb geochronology and isotope geology of the Pan-African Nabitah Orogenic Belt of the Saudi Arabian Shield. In *The Pan African Belt of NE Africa and adjacent areas* (ed. S. El Gaby & R. Greiling), pp. 227–288. Braunschweig: Vieweg.

Wooden, J. L., Stacey, J. S., Howard, K. A., Doe, B. R. & Miller, D. M. 1986 Pb isotopic evidence for the formation of Proterozoic crust in the southwestern United States. In *Metamorphism and crustal evolution of the western United States, Rubey vol. VII* (ed. W. G. Ernst), pp. 68–85. Englewood Cliffs, New Jersey: Prentice-Hall.

Wust, H. J., Reischmann, T., Kroner, A. & Todt, W. 1987 Conflicting Pb–Pb, Sm–Nd and Rb–Sr systematics in Late Precambrian metasediments and metavolcanics from the Eastern Desert of Egypt (abstr). *Terra Cognita* **7**, 334.

Zartman, R. E. & Doe, B. R. 1981 Plumbotectonics – the model. *Tectonophysics* **75**, 135–162.

Discussion

J. R. VAIL (*Portsmouth Polytechnic, U.K.*). The application of geochemical techniques to the classification of geological units within Proterozoic basement areas has been a major factor in distinguishing and characterizing these features. In particular, the sophisticated application of trace element and isotopic ratios for metamorphosed and tectonically disturbed rocks has been most successful in understanding the nature and distribution of geological units. The points raised by the speaker, and the case study of northeastern Africa and adjacent areas are well taken and provide a clear example of how geochemistry can support investigations of the geology of this, and similar, complex areas.

There has been much discussion and some strong criticism on the merits and appropriateness of the terrane concept. There is no doubt, however, that in the Pan-African Afro-Arabian Shield this concept has been useful. It is true that the application so far has been largely geographic, and that names are proliferating and liable to cause confusion, yet in this vast area too little is known of the detailed geology, the age, or the boundary conditions to adopt initially anything other than a simplistic approach. True, the interpretation of boundaries and of dismembered ophiolite remnants as suture zones has been too hasty and will need to be modified as data accumulate; nevertheless the overall regional pattern strongly supports the concept of a continental crustal terrane against a plate margin island-arc oceanic terrane, with

91

internal subterranes all with strongly deformed bounding margins. The paper as presented may have given an erroneous impression as to extent of these features (Vail 1987, 1988).

N. B. W. HARRIS. I agree with Professor Vail that the terrane concept has been useful in interpreting the evolution of the Afro-Arabian shield, although the identification of terrane boundaries through joining up ophiolite fragments across hundreds of kilometres has been, in the main, a fruitless pursuit. On the other hand the identification of continental regions within the island arc collage has been successful, though whether these represent far-travelled terranes or near *in situ* fragments of continental margins remains a matter of speculation.

P. F. HOFFMAN (*Geological Survey of Canada*). Rates of crust formation estimated from Nd model ages tend to be too fast because these ages give only the mean age of crust formation, thereby underestimating the duration of crust-forming episodes.

N. B. W. HARRIS. In the case of juvenile magmas, crystallization ages and model Nd ages are similar. In all other cases mixed sources may contribute to magmatism so that Dr Hoffman's point is a valid one. For the Afro-Arabian shield we have estimated that maximum age span for the *crystallization* of juvenile Pan-African magmas. Hence crustal-growth rates are minimum estimates.

Additional references

Vail, J. R. 1987 Late Proterozoic tectonic terranes in the Arabian-Nubian Shield and their characteristic mineralization. *Geol. J.* **22**, 161–174.
Vail, J. R. 1988 Tectonics and evolution of the Proterozoic basement of northeastern Africa. In *The Pan African Belt of NE Africa and adjacent areas* (ed. S. El Gaby & R. D. Greiling), pp. 195–226. Braunschweig: Vieweg.

Metamorphism in allochthonous and autochthonous terranes of the western United States

By W. G. Ernst

School of Earth Sciences, Mitchell Building 101, *Stanford University, Stanford, California* 94305-2210, *U.S.A.*

The haphazard accretion of exotic terranes during continental reassembly results in a crustal college typified by genetically unrelated lithotectonic belts. Profound chronologic, lithologic, geochemical, and metamorphic breaks characterize such suture zones. However, post-metamorphic differential vertical uplift and erosion can generate a marked discontinuity in grade within a single lithotectonic entity, and in contrast, post-amalgamation recrystallization of an exotic terrane assembly can produce an isofacial metamorphic overprint. Thus the tectonic context of metamorphic mineral parageneses must be interpreted with caution.

In spite of the presence of allochthonous terranes, the western U.S. Cordillera in general is characterized by gradual sectorial enlargements towards the modern edge of the continent, by coherent, broadly continuous isotopic or geochemical provinces, and by systematic oceanward decreases in the metamorphic intensities of the constituent lithic assemblages, both within a belt, and across a series of belts. These relationships hold over a wide range of scales, from that of a physiographic province to that of a quadrangle-sized area. Examples described include chronologic, isotopic, igneous and metamorphic belts of (1) the entire western conterminous U.S. Cordillera, (2) the Phanerozoic Sierran–Klamath basement terrane assembly, and (3) the Great Valley and Franciscan sedimentary couplet derived from the late Mesozoic Sierra Nevada–Klamath arc. For these cases, systematic recrystallization–deformation trends and nearly *in situ* growth of sialic crust are evident. Mapped metamorphic and structural discontinuities reflect dislocations involving spatially associated, co-evolving continental lithotectonic units, and, except for far-travelled oceanic fragments, do not imply wholesale juxtaposition of exotic, genetically unrelated terranes.

Introduction

Differential plate-tectonic motions profoundly affect P–T evolution of the continental crust, and constituent metamorphic assemblages (Miyashiro 1973). Constructional stages are typified by head-on and oblique convergent plate motions, resulting in paired orogenic belts. These comprise: firstly, the development of broad inboard, low P/T, high heat-flow recrystallization régimes characterized by the addition of primary calc-alkaline igneous arc rocks, and the anatectic, metamorphic, and sedimentary reworking of preexisting materials; and secondly, the formation of narrow outboard metamorphic belts of high P/T, low heat-flow subduction complexes which include the suturing of far-travelled oceanic assemblages (with or without island arcs) and allochthonous microcontinental fragments, and are characterized by a lack of coeval calc-alkaline igneous activity. Young paired belts provide a characteristic pattern which documents the nature of geologically recent continental accretion. Episodic rifting and strike–slip faulting, which attend divergent and transform plate motions, modify and obscure the record of sialic growth. Remnants of old metamorphic belts are scattered

93

piecemeal, and annealed or recrystallized by later thermotectonic events. Extension and transform motions thereby rearrange previously produced terranes, but do not increase the aggregate mass of the sialic crust. Of course, these processes do result in the enlargement of acceptor assemblies at the expense of donor continents.

Experimental phase relations for synthetic and natural rock and mineral systems, taken together with $^{18}O/^{16}O$ and D/H isotopic geothermometers and diverse mineralogic thermobarometers, allow the erection of a metamorphic-facies grid (see, for example, Ernst 1976; Liou *et al.* 1985). The thermal structures of divergent and convergent plate boundaries have been modelled numerically by many workers (e.g. Oxburgh & Turcotte 1971). Combining mineralogic *P–T* grids with computed temperature-depth arrangements, the geologic disposition of metamorphic facies for divergent and convergent lithospheric plate junctions have been approximated (Ernst 1976; Zen 1988). The very high heat-flow régime characterized by an oceanic spreading system accounts for a telescoped metamorphic zonation, and relatively shallow-level development of high-rank hornfels beneath the ridge. In contrast, the more complex convergent plate-tectonic setting is characterized by a broad, relatively high heat-flow environment in the magmatic arc, reflected in a crude bilaterally symmetric upwarp of the metamorphic-facies assemblages, in the anatexis of basal portions of thick, 'juicy' continental crust, and in a spectacular downward, asymmetric projection of relatively high P/T phase compatibilities in the subduction zone.

Crustal metamorphic environments and geochemical signatures of igneous rocks derived from the mantle and/or the deep crust, combined with regional and local structural–geologic and lithologic–tectonic age relationships, provide constraints regarding the complex interplay of processes attending continental accretion. Isotopic data suggest that for the conterminous western U.S. Cordillera, the bulk of sialic material was added northwest and, especially, south of the Archaean shield during early and mid-Proterozoic time, whereas late Proterozoic–Phanerozoic growth occurred predominantly along the western margin of the North American craton. Continental enlargement by igneous processes was accompanied by the development of successively younger metamorphic belts. Geologic relations of the Franciscan, Great Valley, and Sierra Nevada–Klamath lithotectonic triad provide a well-preserved example of Phanerozoic continental growth and concomittant metamorphism; it provides a possible analogue for older, inboard lithotectonic complexes. New maps of portions of the central Klamath Mountains and the eastern California Coast Ranges reveal small-scale details of the Mesozoic, nearly in-place accretion of the studied areas. Insight regarding crustal growth and recrystallization may be obtained through an integration of the information obtained from investigations at several scales such as these.

The relationship between metamorphic facies assemblages and terrane accretion is not always straightforward. In favourable cases, pronounced contrasts in grade occur across terrane boundaries, reflecting the post-metamorphic juxtaposition of genetically unrelated lithotectonic units. However, within a single terrane, differential uplift across a late fault can result in the surface exposure of a marked discontinuity in metamorphic grade. On the other hand, post-accretionary recrystallization of an exotic terrane assembly can produce a homogeneous, monotonic metamorphic zonation. Clearly, unambiguous tectonic interpretation of metamorphic belts requires the existence of important chronologic and structural constraints.

METAMORPHISM IN TERRANES OF THE WESTERN U.S.

ARCHAEAN TO CENOZOIC TECTONOMETAMORPHIC BELTS OF THE WESTERN CONTERMINOUS U.S.

Regional metamorphism

Maps summarizing the predominant mineral facies and times of metamorphism in the western Cordillera are presented as figures 1 and 2, respectively. Compilations are based on relationships described by many workers (e.g. Ernst 1988). Although the Mesozoic history of the continental margin is reasonably well known, younger metamorphic tracts are still largely buried. In contrast, pre-Mesozoic belts have been deeply eroded, or overprinted and dislocated by later dynamothermal or tectonic processes, thus are imperfectly preserved. Mesozoic belts

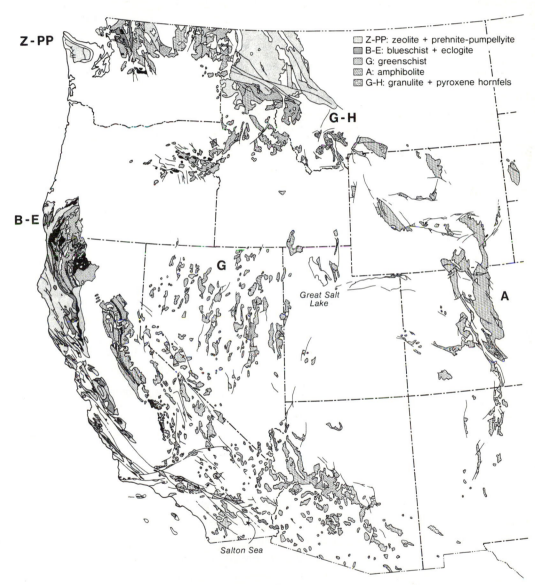

FIGURE 1. Generalized areal disposition of metamorphic belts of the western conterminous U.S., based on regional summary contributions in Ernst (1988); serpentinized peridotites shown in black. The dominant regional metamorphic facies assemblages are presented, irrespective of age. Relict earlier assemblages, and incipient later parageneses are not shown. Some metamorphic facies are combined, as shown in the legend.

95

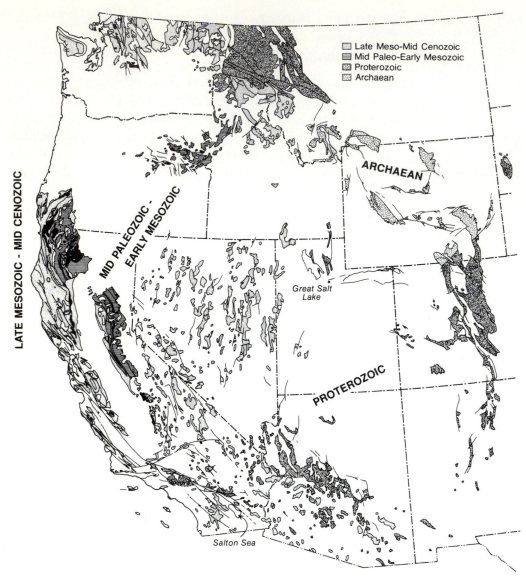

FIGURE 2. Approximate ages of principal recrystallization events characteristic of the dominant metamorphic
assemblages (see figure 1), for metamorphic belts of the western conterminous U.S., based on regional summary
contributions in Ernst (1988); serpentinized peridotites shown in black. Traces of earlier events, and later,
feeble overprintings are not shown. Ages of metamorphism are combined, as shown in the inset.

apparently represent the optimum stage of preservation and exposure for deciphering
petrogenetic evolution of the western U.S.

Archaean metamorphism, of which chiefly high-T, low- to moderate-P amphibolite and
granulite facies assemblages are preserved, typifies the basement of Wyoming and adjacent
parts of Idaho, Utah, Montana and South Dakota. Orogenic activity ceased by late Archaean
time, and much of the ancient crust was not influenced significantly by later dynamothermal
metamorphism. Weak recrystallization characterizes the upper Proterozoic passive-margin
section northwest of the Wyoming nucleus. Intermediate-P amphibolite-type parageneses
developed in mid-Proterozoic accretionary sialic crust to the south in Colorado, Utah and New
Mexico. A regional transition links greenschists in southeastern Arizona to low-P amphibolites

and low-rank granulites in western Arizona. In southeastern California, Proterozoic high-rank, low-P amphibolites are scattered along the San Andreas transform system. To the northwest in California, and in east-central Oregon and northwestern Washington, segments of once more continuous metamorphic belts reflect Ordovician through Cretaceous recrystallization events; sutures within these vestiges of palaeo-Pacific convergent margins are marked by tectonized meta-ophiolites, mantle fragments, and rare allochthonous microcontinental scraps. Oceanward subduction-zone, high-P, non-volcanic metamorphic belts are juxtaposed against landward, high-T, continental-margin, calc-alkaline batholithic realms and pre-existing metamorphic wall rocks.

Westward from the North American platform, lithotectonic belts have been overprinted by late Jurassic–Cretaceous and early-mid Cenozoic intracontinental polymetamorphism accompanying widespread intermediate and silicic igneous activity. Mobilization of inboard, high-rank core complexes (Coney 1980; Armstrong 1982) and broad-scale development of greenschist and low-P amphibolite facies assemblages in the Great Basin, Mojave–Sonoran Desert, and the Sierra Nevada–Peninsular Ranges occurred at this time. Greater depths of emplacement are recorded in country rocks surrounding and north of the Idaho batholith, where high-P amphibolites of late Cretaceous metamorphic age are exposed. Outboard towards the modern North American margin in western California and northwestern Washington, accretionary prisms containing oceanic as well as terrigenous materials have been subjected to zeolite to blueschist facies metamorphism during late Triassic and younger suturing against the continent.

Metamorphic terranes are successively younger from the Wyoming nucleus toward the Pacific Basin and the Gulf of Mexico. Where detailed lithotectonic relationships are available within a province (e.g. Cheyenne belt, Colorado-New Mexico, northwestern Washington, Klamaths, Sierra Nevada, California Coast Ranges), formational ages of original rock assemblages and times of recrystallization decrease seaward. The observed temporal sequence is compatible with nearly *in situ* growth, but would be coincidental if exotic slices of older and younger oceanic and microcontinental terranes had been stranded haphazardly at the accreting western margin of North America.

Paired metamorphic belts of the Cordillera are well preserved only in younger Phanerozoic sections bordering the Pacific Ocean. Inland, recrystallized terranes of diverse ages exhibit lithologic assemblages produced mainly in continental-margin and island-arc settings. Blueschist belts and extensive tracts of peridotite are lacking; if such associations were formed in early Palaeozoic and older plate-tectonic environments of the western U.S., they must have been selectively destroyed by subsequent thermotectonic events or transported away along margin-parallel shear zones.

Isotopic data

Tectonometamorphic trends in the western Cordillera are clearest for the Phanerozoic complexes. Summary of the isotopic data for sialic igneous rocks helps to illuminate the more obscure crustal growth and petrogenesis of less well-preserved Precambrian metamorphic belts. Melts derived from partial fusion of deep-crustal and upper-mantle source materials provide geochemical constraints on the nature of the basement. Mesozoic and Cenozoic magmatic rocks reflect a continental lithosphere typified by discrete Pb and Nd isotopic provinces (Zartman 1974; Farmer & DePaolo 1983; Wooden *et al.* 1988); these igneous units exhibit variable contributions from mantle and crustal protoliths. The age of separation of the continental crust from an evolving mantle reservoir, as indicated by Precambrian Pb and Nd model and crystallization ages, monotonically decreases from the Archaean Wyoming province through the early Proterozoic central Great Basin and Mojave Desert to the mid-Proterozoic Sonoran Desert (Bennett & DePaolo 1987). Within the Mesozoic calc-alkaline belts, a systematic oceanward decrease in the degree of continental involvement, as expressed by decreasing $^{87}Sr:^{86}Sr$ initial ratios, and increasing ϵ_{Nd} values for volcanics and plutonics, is well documented (Kistler & Peterman 1978; DePaolo 1981; Condie 1986; Farmer & DePaolo 1983; Bickford 1988).

Accretion and accompanying recrystallization of the western U.S. continental crust occurred principally during three major intervals, through the primary formation of calc-alkaline batholiths and superjacent volcanogenic arcs: (1) the later Archaean (2.5–3.3 Ga), when the continental nucleus was assembled; (2) the early and mid-Proterozoic (1.4–2.3, chiefly 1.7–1.9 Ga), when most of the sialic basement was generated progressively southward; and (3) the

FIGURE 3. Isotopic provinces of the western conterminous U.S. based on Nd-depleted model mantle ages (Farmer & DePaolo 1983; Farmer 1988; Bennett & DePaolo 1987), and $^{87}Sr:^{86}Sr$ ($= 0.706$ and 0.704) initial ratio limits in Mesozoic and Cenozoic granitoids (Armstrong *et al.* 1977; Kistler & Peterman 1978). Mid-Proterozoic and older basement lies inboard of the 0.706 line. Lead isotopic provinces (Zartman 1974) exhibit analogous areal dispositions.

Phanerozoic, when the continent grew westward. The age of separation from the mantle of materials constituting the Precambrian basement gradually decreases from the Wyoming province (greater than 2.6 Ga), through the central Great Basin and Mojave Desert (2.0–2.3 Ga) and the Colorado Plateau (1.8–2.0 Ga), to the Sonoran Desert (1.7–1.8 Ga), and the mid-continent granite-rhyolite belt (less than 1.4 Ga). Igneous crystallization ages and times of metamorphism mirror these trends.

Geochemical–radiometric provinces are shown in figure 3. Isotopically distinct belts border the Archaean craton, and exhibit overall Proterozoic and Phanerozoic enlargements towards the modern continental margins. Relationships hold for model mantle separation ages as well as times of igneous crustal formation. Lithotectonic belts apparently developed asymmetrically and episodically rather than concentrically and continuously, indicating sequential lateral growth rather than continuous circumferential enlargement. They evidently mark the sites of subduction zones and landward arcs. Truncations of these belts suggest the episodic rifting and removal of gradually accreted segments of the North American continental crust (Burchfiel & Davis 1975).

Broad trends typifying Precambrian crustal evolution and coeval metamorphism are compatible with overall growth of the western Cordillera mainly by the generation and telescoping of new continental crust surmounting convergent plate junctions, with the incorporation of variable amounts of recycled sialic material, and by the random accretion of exotic, mainly oceanic terranes of unrelated geochemistry. Because of the observed systematic decrease in igneous and metamorphic ages, and coherence of isotopic data for the belts proceeding outward from the Wyoming craton (greater than 2.6 Ga), it is evident that addition of far-travelled microcontinental terranes consisting of sialic basement detectably older than the developing margin did not characterize growth of the western conterminous U.S. Possible exceptions to this generalization include: the Salinian granitic salient west of the San Andreas fault (Hill & Dibblee 1953; Page 1982); easternmost California, where a westward step in the $^{87}Sr:^{86}Sr$ initial ratio offsetting the late Precambrian continental margin may mark a faulted terrane boundary (Kistler & Peterman 1978); and the Yellow Aster Complex (Misch 1966) of the northwestern Cascades, a small, outboard fragment of old sialic crust.

SIERRA NEVADA–KLAMATH BASEMENT COMPLEX

The Sierra Nevada appears to represent the southeastern continuation of the Klamath province (Davis 1969; Day *et al.* 1988). The Klamaths consist of metamorphosed country rocks intruded by isolated, pre-Cretaceous calc-alkaline plutons, whereas the Sierra Nevada consists principally of coalescing, Jurassic–Cretaceous batholithic units, separated incompletely by thin metamorphic septa. The northwestern foothills belt is the only major segment of Sierran wall rocks preserved adjacent to the plutonic series, although isolated roof pendants occur throughout the range. Compared with the Klamaths, the Sierra Nevada may represent a deeper level of crustal exposure, especially at its southern extremity where high-rank amphibolitic or low-rank granulitic meta-igneous rocks are exposed beneath the batholith (Sams & Saleeby 1988). The dispositions of metamorphic belts and lithotectonic terranes of northern and central California are presented in figures 4 and 5, respectively.

Sutures bounding individual Sierran and Klamath metamorphic belts are marked by

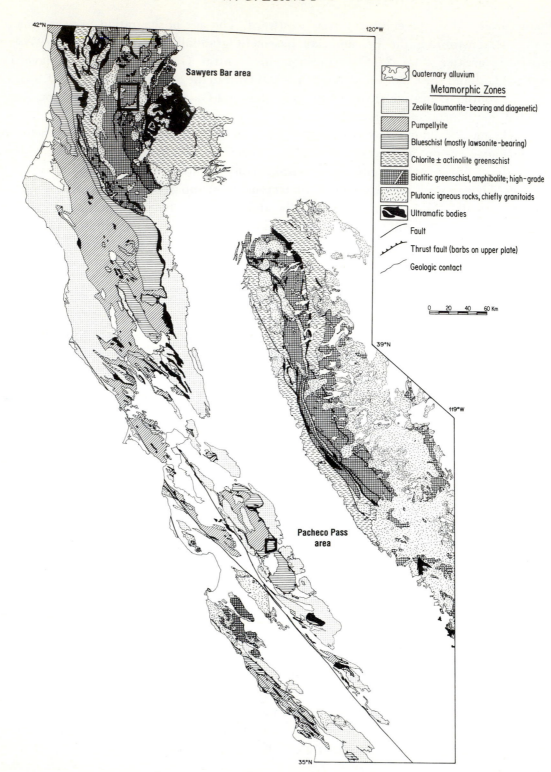

FIGURE 4. Metamorphic zonations developed in lithotectonic belts of northern and central California, based on the geologic map of California (Jennings 1977), and generalized by Ernst (1983) from numerous literature sources. High-grade metamorphic rocks are present, especially in Sierran roof pendants, in the Klamath central metamorphic belt, and in the Salinian terrane. Locations of the Sawyers Bar area (figure 6) and the Pacheco Pass area (figure 8) are indicated.

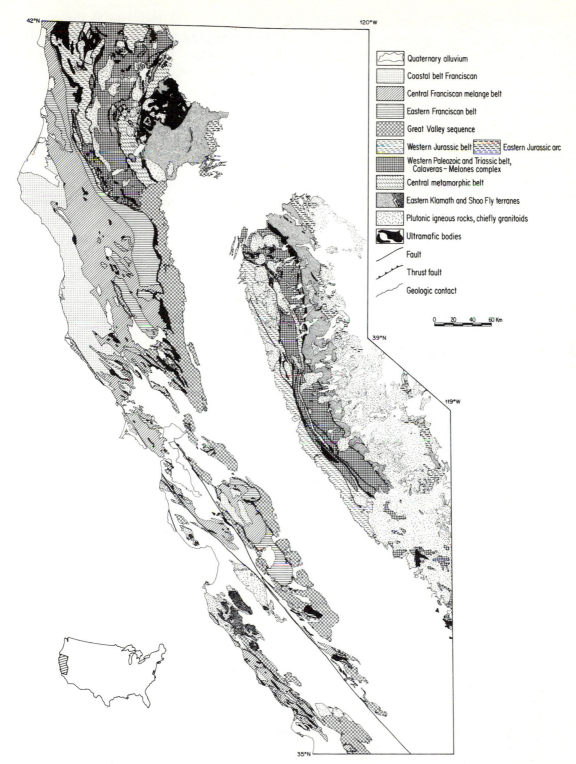

Legend:
- Quaternary alluvium
- Coastal belt Franciscan
- Central Franciscan melange belt
- Eastern Franciscan belt
- Great Valley sequence
- Western Jurassic belt / Eastern Jurassic arc
- Western Paleozoic and Triassic belt, Calaveras – Melones complex
- Central metamorphic belt
- Eastern Klamath and Shoo Fly terranes
- Plutonic igneous rocks, chiefly granitoids
- Ultramafic bodies
- Fault
- Thrust fault
- Geologic contact

0   20   40   60 Km

FIGURE 5. Lithotectonic belts of northern and central California based on the geologic map of California (Jennings 1977), and generalized by Ernst (1983) from numerous literature sources. The eastern, mid-Mesozoic sedimentary and volcanic stratified rocks of the eastern Klamath and northern Sierra are depicted with a map pattern similar to that of the western Jurassic belt because, although not necessarily related, these sections were deposited nearly contemporaneously.

101

serpentinized peridotites, but the amount of ultramafic material decreases to the southeast. Different lithotectonic units (Irwin 1981; Sharp 1988) from east to west include: the early-to-mid-Paleozoic Shoo Fly and eastern Klamath terranes; the Feather River and Trinity peridotites; the central metamorphic belt (principally in the Klamaths); blueschists of the Melones fault zone and the Stuart Fork Formation; the late-Paleozoic–early-Mesozoic Sierran Calaveras–Melones complex and so-called western Triassic and Paleozoic belt of the Klamath Mountains; and the western Jurassic belt. These lithotectonic units display east-dipping imbrication; isoclinally folded, west-vergent sections have been recognized, but where primary flow tops and sedimentary laminations are preserved, sections typically face east (Clark 1964, 1976). Many of the individual belts may represent terrane assemblies (Irwin 1972; Silberling *et al.* 1987). In general, age of formation, metamorphic grade and structural complexity all

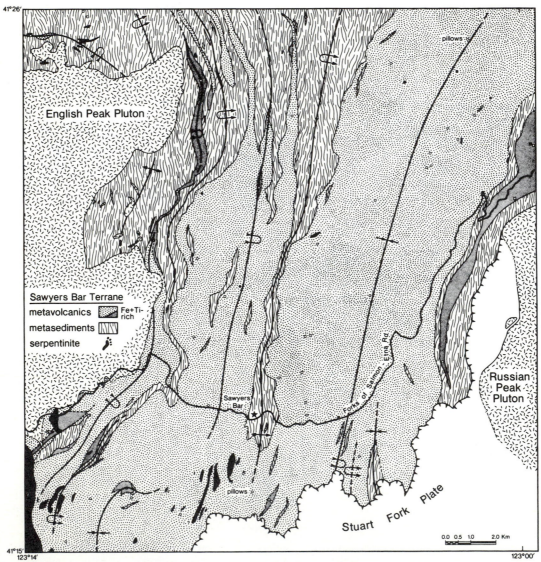

FIGURE 6. Geology of the Sawyers Bar area, central Klamath Mountains (after Ernst (1987) with additional mapping). See figure 4 for location. The intricately interfingered metasedimentary units of the more westerly (equivalent to Hayfork) and easterly (equivalent to North Fork and Salmon River) terranes are part of a single complex. The town of Sawyers Bar is indicated by a black star.

increase eastward (Schweickert 1981; Schweickert *et al.* 1988). Outboard units contain terrigenous debris and tectonic blocks derived from the more landward terranes (Behrman & Parkison 1978; Wright 1982; Ando *et al.* 1983; Gray 1986; Hannah & Moores 1986), suggesting that most terranes are native to the North American margin.

As an example, detailed mapping of greenschist-facies rocks in the central part of the western Triassic and Palaeozoic belt of the central Klamath Mountains near Sawyers Bar, California, demonstrates stratal continuity and the interlayered nature of intricately folded, distinctive metabasaltic and metaturbiditic units. Geologic relationships are presented in figure 6. Although metavolcanic and metasedimentary units of this area were previously regarded as portions of several disparate terranes (Donato 1987; Silberling *et al.* 1987), these supracrustal rocks all belong to a single lithostratigraphic entity. Contacts in the mapped area are depositional rather than faulted, except for the east-dipping thrust fault bringing the Stuart Fork plate over Sawyers Bar rocks. Petrologic, structural, geologic, and geochemical studies demonstrate the absence of metamorphic, tectonic, stratigraphic, and bulk-rock compositional discontinuities. Evidently the Sawyers Bar segment of the western Triassic and Paleozoic belt formed and evolved as a single, possibly nearly in-place arc, rather than as a collage of unrelated, exotic crustal entities.

Within any one Sierran–Klamath complex, recrystallization intensity in general decreases oceanward, as does the age of formation, and this progression tends to hold on a grander scale for the various juxtaposed belts (Ernst 1983). For example, in the Klamaths, metamorphic grade decreases westward from upper amphibolite facies in the central metamorphic belt through greenschist facies to prehnite-pumpellyite facies in the western Jurassic belt (Irwin 1981; Burchfiel & Davis 1981; Coleman *et al.* 1988). The age of origin for the protoliths decreases commensurately from mid-Palaeozoic to late Jurassic. A comparable metamorphic–temporal regional arrangement exists in the northwestern Sierran foothills (Saleeby 1981; Day *et al.* 1988).

NORTHERN AND CENTRAL FRANCISCAN–GREAT VALLEY DEPOSITIONAL COMPLEX

Franciscan and Great Valley units are areally associated from southern Oregon to west-central Baja California. Both consist dominantly of clastic sedimentary rocks, recrystallized to contrasting extents, and each has a latest Jurassic–Palaeogene depositional range. The ensimatic, tectonically imbricated prism of partly chaotic (Hsü 1968), chiefly east-dipping and east-facing Franciscan strata, lies closest to the Pacific Ocean and is confined to the Coast Range province; it is separated from the broadly contemporaneous Klamath–Sierran igneous arc by the well-bedded Great Valley Group. The two terrigenous sedimentary belts have been regarded as genetically related trench and forearc-basic deposits, respectively (Ernst 1970; Dickinson 1972, 1976).

However, palaeomagnetic and fossil data document the far-travelled nature of Franciscan deep-sea cherts and pillow basalts (see, for example, Blake 1984; Jayko & Blake 1984; Beck 1986). These facts have evoked the hypothesis of juxtaposition of unrelated Franciscan and Great Valley terranes (Blake & Jones 1978; Blake *et al.* 1988), compatible with the accretion of exotic terranes throughout the western Cordillera (Coney *et al.* 1980; Jones *et al.* 1983). The present spatial association of the two metasedimentary assemblages in the California Coast Ranges reflects relative westward thrusting of Great Valley strata and underlying Coast Range ophiolite over the high-*P* coeval Franciscan Complex (Bailey *et al.* 1964, 1970), followed by, or concurrent with, return flow for the subducted Franciscan (Suppe 1972; Cloos 1982; Platt 1986). Seismic reflection and refraction profiles indicate that the Franciscan may have wedged eastward between overlying Great Valley strata and the Sierran basement (Wentworth *et al.* 1984); however, inasmuch as Quaternary units are involved, this may represent chiefly Neogene deformation.

Although most rock types of oceanic character are exotic, geologic relations suggest that the volumetrically dominant clastic rocks of western California are native to the North American margin. Studies of paleocurrent vectors and sandstone plus conglomerate petrofacies demonstrate a common Klamath–Sierran provenance for Franciscan and Great Valley detritus (Telleen 1977; Jacobson 1978; Dickinson *et al.* 1982; Ingersoll 1979; Seiders & Blome 1984; Seiders 1988). Similarities of quartz, feldspar, and lithic-fragment proportions within the forearc basic and trench complex, inferred palaeogeography, and sediment-distribution trajectories are illustrated in figure 7. Spatial contiguity seems to be required during deposition of these units. Moreover, Upper Cretaceous Great Valley trench-slope units locally rest with angular unconformity on the underlying Franciscan (Maxwell 1974; Smith *et al.* 1979). Forearc deposition evidently took place in intimate proximity to a rising mass of decoupled trench melange.

Four major lithotectonic belts, divided by some workers into more numerous tectono-stratigraphic terranes, crop out in the northern Coast Ranges, and on a smaller scale, in

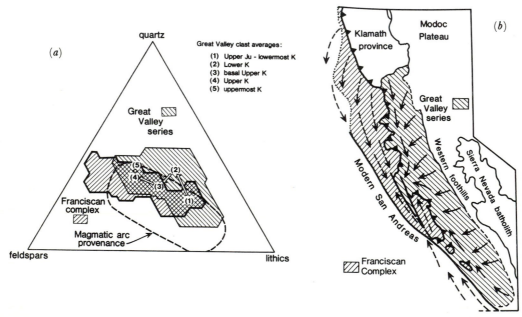

FIGURE 7. (*a*) Modal proportions of clastic quartz, feldspars and lithic fragments from 232 Great Valley and 203 Franciscan sandstones from northern and central California, compared with typical magmatic-arc derived sandstones, after Dickinson *et al.* (1982). Evolution in average Great Valley modes during unroofing of volcanic and metamorphic cover, and exposure of K-feldspar-bearing granitoids is illustrated by the petrostratigraphic intervals (1)–(5). Similarities in Great Valley and Franciscan conglomerate clasts have been documented by Seiders (1988). (*b*) Cretaceous palaeogeography and sediment distribution paths in northern and central California, after Dickinson *et al.* (1982). The major suture zone shown as a thrust contact with barbs on the upper plate is the Coast Range fault, which experienced compound movement, including earlier subduction (underflow and compression), and later uplift (return flow and extension) as documented by Bailey *et al.* (1970), Ernst (1970), Cloos (1982), Platt (1986), and Jayko *et al.* (1987).

the southern Coast Ranges (Blake *et al.* 1985, 1988). From east to west, these are: (*a*) the Great Valley Group; (*b*) the eastern Franciscan belt; (*c*) the central Franciscan melange; and (*d*) the coastal belt of Franciscan. Great Valley strata rest unconformably on inboard Sierran–Klamath continental basement, and farther to the west, on the Coast Range ophiolite. Franciscan sediments were laid down on an unknown substrate, but mafic tectonic blocks present in the disrupted complex may represent outboard palaeo-Pacific oceanic crust.

(*a*) The well-bedded Great Valley accretionary prism constitutes an asymmetric synclinorium with a near-vertical western limb and gently west-dipping eastern limb (Hackel 1966). Clastic strata are feebly recrystallized to zeolite facies assemblages near the base of the thickest, westerly sections (Dickinson *et al.* 1969; Bailey & Jones 1973). Of similar metamorphic grade, the Coast Range ophiolite crops out discontinuously along the west side of the Great Valley (Bailey *et al.* 1970). It includes several geochemically different mafic–ultramafic units (Shervais & Kimbrough 1985). Some were generated *ca.* 165 Ma ago at an oceanic-spreading centre (Hopson *et al.* 1981), but others are locally associated with slightly younger (*ca.* 153 Ma) calc-alkaline arc rocks (Evarts 1977); most segments of the ophiolite are capped by Upper Jurassic deep-sea radiolarian cherts which pass upwards to rhythmically layered, thin-bedded Great Valley distal mudstones. Ophiolite overlain by magmatic-arc rocks may have been generated near the continental margin, whereas other occurrences, having remnant palaeomagnetic inclinations compatible with near-equatorial formation (Luyendyk 1982), probably are exotic.

(*b*) The eastern Franciscan belt and its southern extension in the Diablo Range consist of a relatively well-ordered series of tectonically imbricated phyllitic schists, quartzofeldspathic metagraywackes, dark metashales, pods of greenstone plus serpentinite, and widespread, thin chert layers. The complex is bounded to the east by the Coast Range fault and the structurally overlying ophiolite and Great Valley Group, to the west by a fault juxtaposing the tectonically lower central Franciscan melange. Sections within the eastern belt, as well as bounding faults, dip eastward (Suppe 1973). Depositional ages of turbiditic strata range from latest Jurassic to mid Cretaceous.

Rocks of the eastern belt are characterized by minor but ubiquitous neoblastic lawsonite. Two distinct lithostratigraphic assemblages make up this terrane: (i) Interlayered schistose metapelitic and metabasaltic blueschists occur in the northeastern part of the Franciscan Complex directly beneath the Coast Range fault (Blake *et al.* 1967; Irwin *et al.* 1974; Brown & Ghent 1983). This entity is the highest structural unit in a stack of east-dipping sheets (Worrall 1981). The minimum metamorphic K/Ar age of these schists is about 120 Ma (Lanphere *et al.* 1978). (ii) Relatively undeformed metagraywacke and metashale sequences, especially extensive along the east side of the Diablo Range, contain widespread metamorphic aragonite and associated jadeitic clinopyroxene+quartz (McKee 1962*a*, *b*; Ernst *et al.* 1970). The time of high-P, low-T recrystallization seems to have been about 90–120 Ma (Suppe & Armstrong 1972; Mattinson & Echeverria 1980).

A geologic map of the Pacheco Pass area of the east-central Diablo Range is shown in figure 8 as a representative area. Units dip predominantly to the east. The largely (meta-)detrital section, which consists of neoblastic jadeitic-pyroxene bearing graywackes and dark shale laminae possesses stratal coherence, judging by the lateral continuity of interbedded cherty layers. Pods of blueschist, greenstone, and minor serpentinite occur particularly in unit I, the stratigraphically lowest part of the metaclastic sequence. This unit in part may represent an olistostrome or tectonic melange. Mafic, chiefly intrusive meta-igneous rocks are rare in the stratigraphically higher members. Thickening and thinning, especially in unit II, is evident. The entire section, regarded by some as a tectonic aggregation of unrelated terranes (see, for example, Blake 1984) appears to be a coherent stratigraphic–metamorphic entity.

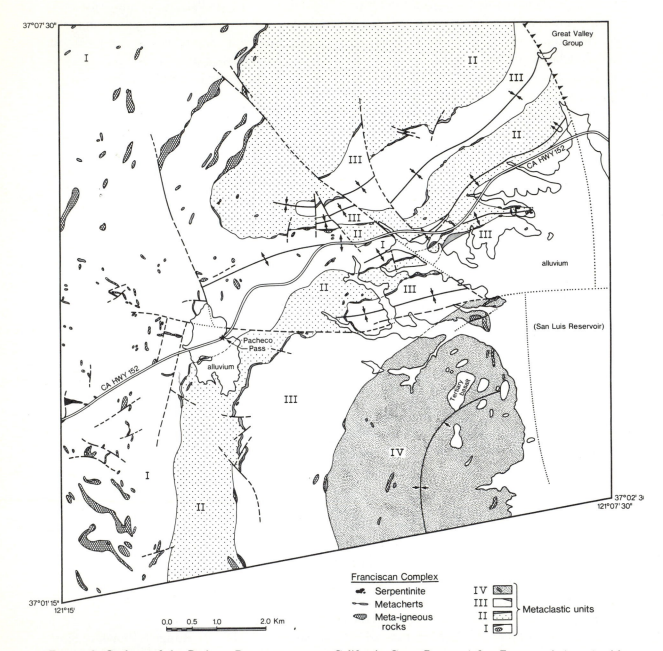

FIGURE 8. Geology of the Pacheco Pass area, eastern California Coast Ranges (after Ernst *et al.* (1970) with additional mapping). See figure 4 for location. Neoblastic jadeitic pyroxene is widespread in quartzose metagraywackes of this region (McKee 1962*a*; Ernst *et al.* 1970; Maruyama *et al.* 1985). The San Luis Reservoir, which covers the alluvium in the eastern part of the map, is not shown. Four local, stratigraphically coherent metaclastic units of this portion of the Franciscan Complex, referred to as units I–IV, are distinguished in this area. The lowermost, unit I, is rich in metavolcanic pods, and may be olistostromal. Stratal continuity is preserved in this part of the belt. Pacheco Pass is indicated by a black star.

(*c*) The central Franciscan belt consists mainly of chaotic melange, with variable proportions of more coherent strata. The sedimentary age of the shaly matrix appears to be mid and late Cretaceous, although included blocks and broken-formation slabs range in age from late Jurassic to late Cretaceous, and are of both foreign and cognate lithologies (Blake & Jones 1978). Fractured and attenuated blocks may be regarded as boudins, whereas with lesser degrees of disruption, tectonic melange passes gradually into semistratified broken formation (Hsü 1968, 1974). The lower the sandstone:shale ratio, the more thoroughly disrupted the sections appear to be. Laminar bodily flow within the subducted melange was modelled by Cowan & Silling (1978) and Cloos (1982). Foreign, previously metamorphosed eclogitic and amphibolitic blocks evidently were spalled off from different loci along boundaries of the subduction-induced circulation system; this phenomenon, coupled with density-induced differential settling velocities of high-grade blocks during flow, accounts for both the lithologic variation and size distribution of the tectonic fragments within the melange (Cloos 1982, 1986).

Metamorphism of the central Franciscan belt at modest *P* and *T* produced pumpellyite and lawsonite, but not jadeitic pyroxene (Blake *et al.* 1967; Suppe 1973; Cloos 1983). The recrystallization age of the matrix is probably late Cretaceous, judging by the contained mid-Cretaceous fossils. Mafic tectonic blocks of garnet-bearing blueschist, amphibolite and eclogite represent fragments of previously subducted oceanic crust, now engulfed in the melange. Although present in the eastern belt, they are most abundant in the central belt. Peripheral actinolite + chlorite or talc rinds indicate that the blocks were not in chemical equilibrium with the enclosing lower metamorphic grade matrix of the olistostrome, tectonic melange or serpentinite host. Recrystallization ages of these high-grade blocks typically are on the order of 160–163 Ma (Coleman & Lanphere 1971; Mattinson 1988), considerably older than the sparsely fossilferous, chaotic matrix in which they are found.

(*d*) The coastal Franciscan belt is bounded on the east by the steeply east-dipping coastal belt thrust, which separates it from the structurally higher central Franciscan melange. Lesser degrees of stratal disruption are characteristic of this rather arkosic Upper Cretaceous–Miocene lithotectonic unit (McLaughlin *et al.* 1982). Premetamorphic rock types include deep-water clastic strata on the west, apparently associated with an altered basaltic substrate and chert, passing eastward by degrees to mid-fan turbiditic, andesitic graywacke and abundant quartzofeldspathic graywacke (Blake & Jones 1978). Greenstone and serpentinite lenses are scarce, and high-grade tectonic blocks are extremely rare in this belt (Blake *et al.* 1988).

Metamorphism in the coastal belt is weak, and has not been investigated extensively. Laumontite seems to be widely distributed. A few metasedimentary parageneses include prehnite and/or pumpellyite, but *in situ* blueschists have not been reported (Bailey *et al.* 1964; McLaughlin *et al.* 1982). Phase assemblages are similar to those described from most deeply buried portions of the Great Valley Group. This seaward belt of Franciscan apparently was not subjected to profound underflow before decoupling from the subducting plate and underplating along the continental margin.

Subduction, transform motion and metamorphic belts of the western U.S. Cordillera

Post-Middle Jurassic sea-floor spreading is recorded in oceanic crust and overlying hemipelagic sediments of the Pacific Basin (Pitman *et al.* 1974; Engebretson *et al.* 1985; Debiche *et al.* 1987). Whereas parts of western limbs of several oceanic plates are still extant, eastern limbs of these spreading systems have been overridden by the North American plate during the past 165 Ma (Hamilton 1969). More than 10 000 km of eastward subduction must have occurred (Ernst 1984), averaging about 6 cm a^{-1}. Thus, in spite of several thousand kilometres of northward drift of oceanic crust-capped lithosphere relative to North America, underflow was the dominant mechanism responsible for production of paired metamorphic belts and voluminous calc-alkaline igneous activity, reflecting Cordilleran continental growth and recrystallization during mid-Mesozoic to mid-Cenozoic time.

Based on measured high heat flow in the California Coast Ranges, fission track analyses, and reasonable mineralogic transformation rates, Dumitru (1989) calculated that rapid thermal obliteration of high-P metamorphic mineral assemblages is currently in progress at crustal depths greater than 5–10 km. The strike–slip régime of western California, therefore, evidently has been recently imposed on a chiefly convergent Mesozoic–Cenozoic continental margin (Atwater 1970); until recently, the process of subduction has sustained the nearly continuous refrigeration and preservation of blueschist lithologies in the Coast Ranges since their late Mesozoic formation. Peacock (1988) drew the same general conclusion to account for inverted metamorphic gradients in the westernmost Cordillera. Later terrane shuffling along the dextral shear system of the California Coast Ranges, therefore, has partly obscured the effects of the main constructional stage of sialic growth and metamorphism accompanying lithospheric plate descent.

Summary

Both the eastward underflow of great tracts of Palaeo-pacific oceanic lithosphere, and the apparent recent change in thermal structure of the westernmost continental crust argue for long-continued subduction as the chief plate tectonic process operating in the western U.S. Cordillera during mid-Mesozoic to mid-Cenozoic time. This mechanism explains the contemporaniety, spatial association, and contrasting P–T histories of landward calc-alkaline arcs, forearc-basin deposits, and oceanward trench complexes. The total absence of old blueschist belts within the interior of the U.S. Cordillera probably is due to thermal overprinting. The scarcity of negatively buoyant ophiolitic peridotites may reflect systematic downward sagging of dense mantle and mafic crustal material accompanying thermal softening and crustal remobilization.

True continental growth resulting from the separation of alkali- and Si-rich material from the mantle, in contrast to rearrangement of fragments of already extant sialic crust, requires an important component of convergent plate motion (Ernst 1984). Orderly Proterozoic and Phanerozoic metamorphism, the Precambrian isotopic provinces and rock record, the gradual oceanward decrease in initial ^{87}Sr:^{86}Sr ratios and increase in ϵ_{Nd} values in continental igneous rocks (see, for example, DePaolo 1981), and both large- and small-scale geologic relationships in Californian terranes all are compatible with nearly *in situ* crustal growth and metamorphism. Near the Pacific margin of the U.S., allochthonous ophiolitic debris is abundant, and exotic

Phanerozoic microcontinental fragments of uncertain source are present but rare. However, accretion primarily involved the sweeping back into the North American margin of native terrigenous debris and previously metamorphosed sialic fragments of local or regional provenance. Development of the late Mesozoic triad, Franciscan trench/Great Valley forearc/Sierran-Klamath calc-alkaline arc provides a relatively well-preserved example of the growth and reworking process; the nearly *in situ* setting is demonstrated by sedimentary provenance and transport vectors as well as by systematic metamorphic, isotopic, and petrochemical trends. Lateral rearrangement of genetically related slices of the margin, as well as occasional removal of continental-margin–island-arc sections and docking at distant sites are important complications. Nevertheless, subduction-related arc processes, and recycling in forearc, backarc, and trench environments, evidently have been responsible for most of the enlargement and attendant metamorphism of continental crust in the western U.S. Cordillera.

I acknowledge the help of participants at a UCLA Rubey Colloquium on 'Metamorphism and crustal evolution of the western United States' during January, 1986. The resultant publication (Ernst 1988), provided the framework for this paper. Support was provided by UCLA, Stanford, and the Department of Energy through grant FG03-87ER13806. This paper has been reviewed and improved by W. R. Dickinson and E-an Zen.

REFERENCES

Ando, C. J., Irwin, W. P., Jones, D. L. & Saleeby, J. B. 1983 The ophiolitic North Fork terrane in the Salmon River region, central Klamath Mountains, California. *Geol. Soc. Am. Bull.* **94**, 236–252.

Armstrong, R. L. 1982 Cordilleran metamorphic core complexes – from Arizona to southern Canada. *A. Rev. Earth planet. Sci.* **10**, 129–154.

Armstrong, R. L., Taubeneck, W. H. & Hales, P. O. 1977 Rb–Sr and K–Ar geochronometry of Mesozoic granitic rocks and their Sr isotopic composition, Oregon, Washington, and Idaho. *Geol. Soc. Am. Bull.* **88**, 397–411.

Atwater, T. 1970 Implications of plate tectonics for the Cenozoic tectonic evolution of western North America. *Geol. Soc. Am. Bull.* **81**, 3513–3536.

Bailey, E. H. & Jones, D. L. 1973 Metamorphic facies indicated by vein minerals in basal beds of the Great Valley Sequence, northern California. *J. Res. (U.S. Geol. Surv.)* **1**, 383–385.

Bailey, E. H., Irwin, W. P. & Jones, D. L. 1964 Franciscan and related rocks, and their significance in the geology of western California. *Calif. Div. Mines Geol., Bull.* **183**, 171 pp.

Bailey, E. H., Irwin, W. P. & Jones, D. L. 1970 On-land Mesozoic oceanic crust in California Coast Ranges. *U.S. Geol. Surv. Prof. Paper* **700-C**, 70–81.

Beck, M. E. 1986 Model for late Mesozoic–early Tertiary tectonics of coastal California and western Mexico and speculation on the origin of San Andreas fault. *Tectonics* **5**, 49–64.

Behrman, P. G. & Parkison, G. A. 1978 Paleogeographic significance of the Callovian to Kimmeridgian strata central Sierra Nevada foothills, California. In *Mesozoic paleogeography of the western United States* (ed. D. G. Howell & K. A. McDougall), pp. 349–360. Pacific Section, Soc. Econ. Paleont. Mineral., Pacific Coast Paleogeography Symp. **2**.

Bennett, V. C. & DePaolo, D. J. 1987 Proterozoic crustal history of the western United States as determined by Neodymium isotopic mapping. *Geol. Soc. Am. Bull.* **99**, 674–685.

Bickford, M. E. 1988 The accretion of Proterozoic crust in Colorado: igneous sedimentary, deformational, and metamorphic history. In *Metamorphism and crustal evolution of the western United States* (ed. W. G. Ernst), pp. 411–430. Englewood Cliffs, New Jersey: Prentice-Hall.

Blake, M. C. Jr (ed.) 1984 Franciscan Geology of Northern California. *Pac. Sec., Soc. Econ. Paleont. Mineral.* **43**, 254.

Blake, M. C. Jr, Irwin, W. P. & Coleman, R. G. 1967 Upside-down metamorphic zonation, blueschist facies, along a regional thrust in California and Oregon. *U.S. Geol. Surv. Prof. Paper* **575-C**, 1–9.

Blake, M. C. Jr, Jayko, A. S. & McLaughlin, R. J. 1985 Tectonostratigraphic terranes of the northern Coast Ranges, California. In *Tectonostratigraphic terranes of the circum–Pacific region* (ed. D. G. Howell), pp. 159–171. Circum–Pacific Council Energy Mineral Resources, Earth Science Serial no. **1**.

Blake, M. C. Jr, Jakyo, A. S., McLaughlin, R. J. & Underwood, M. B. 1988 Metamorphic and tectonic evolution of the Franciscan Complex, northern California. In *Metamorphism and crustal evolution of the western United States* (ed. W. G. Ernst), pp. 1035–1060. Englewood Cliffs, New Jersey: Prentice-Hall.

Blake, M. C. Jr & Jones, D. L. 1978 Allochthononous terranes in northern California? A reinterpretation: In *Mesozoic paleography of the western United States* (ed. D. G. Howell & K. A. McDougall), pp. 397–400. Pacific Section, Soc. Econ. Paleont. Mineral., Pacific Coast Paleogeography Symposium **2**.

Brown, E. H. & Ghent, E. D. 1983 Mineralogy and phase relations in the blueschist facies of the Black Butte and Ball Rock areas, northern California Coast Ranges. *Am. Miner.* **658**, 365–372.

Burchfiel, B. C. & Davis, G. A. 1975 Nature and controls of Cordilleran orogenesis, western United States: extensions of an earlier synthesis. *Am. J. Sci.* A **275**, 363–396.

Burchfiel, B. C. & Davis, G. A. 1981 Triassic and Jurassic tectonic evolution of the Klamath Mountains–Sierra Nevada geologic terrane. In *The geotectonic development of California* (ed. W. G. Ernst), pp. 50–70. Englewood Cliffs, New Jersey: Prentice-Hall.

Clark, L. D. 1964 Stratigraphy and structure of part of the western Sierra Nevada metamorphic belt, California. *U.S. Geol. Surv. Prof. Paper* **410**.

Clark, L. D. 1976 Stratigraphy of the north half of the western Sierra Nevada metamorphic belt, California. *U.S. Geol. Surv. Prof. Paper* **923**.

Cloos, M. 1982 Flow melanges: numerical modeling of geological constraints on their origin in the Franciscan subduction complex, California. *Geol. Soc. Am. Bull.* **93**, 330–345.

Cloos, M. 1983 Comparative study of melange matrix and metashales from the Franciscan subduction complex with the basal Great Valley sequence, California *J. Geol.* **91**, 291–306.

Cloos, M. 1986 Blueschists in the Franciscan Complex of California: petrotectonic constraints on uplift mechanisms. *Geol. Soc. Am. Mem.* **164**, 77–93.

Coleman, R. G. & Lanphere, M. A. 1971 Distribution and age of high-grade blueschists, associated eclogites, and amphibolites from Oregon and California. *Geol. Soc. Am. Bull.* **82**, 2397–2412.

Coleman, R. G., Mortimer, N., Donato, M. M., Manning, C. E. & Hills, L. B. 1988 Tectonic and regional metamorphic framework of the Klamath Mountains and adjacent Coast Ranges, California and Oregon. In *Metamorphism and crustal evolution of the western United States* (ed. W. G. Ernst), pp. 1061–1097. Englewood Cliffs, New Jersey: Prentice-Hall.

Condie, K. C. 1986 Geochemistry and tectonic setting of early Proterozoic supracrustal rocks in the southwestern United States. *J. Geol.* **94**, 845–864.

Coney, P. J. 1980 Cordilleran metamorphic core complexes: an overview. In *Cordilleran metamorphic core complexes* (ed. M. D. Crittenden Jr, P. J. Coney & G. H. Davis), pp. 7–31. *Geol. Soc. Am. Mem.* **153**.

Coney, P. J., Jones, D. L. & Monger, J. W. H. 1980 Cordilleran suspect terranes. *Nature, Lond.* **288**, 329–333.

Cowan, D. S. & Silling, R. M. 1978 A dynamic, scaled model of accretion at trenches and its implications for the tectonic evolution of subduction complexes. *J. geophys. Res.* **83**, 5389–5396.

Davis, G. A. 1969 Tectonic cotrelations, Klamath Mountains and western Sierra Nevada, California. *Geol. Soc. Am. Bull.* **80**, 1095–1108.

Day, H. W., Schiffman, P. & Moores, E. M. 1988 Metamorphism and tectonics of the northern Sierra Nevada. In *Metamorphism and crustal evolution of the western United States* (ed. W. G. Ernst), pp. 737–763. Englewood Cliffs, New Jersey: Prentice-Hall.

Debiche, M. G., Cox, A. & Engebretson, D. 1987 The motion of allochthonous terranes across the North Pacific Basin. *Geol. Soc. Am. Spec. Paper* **207**.

DePaolo, D. J. 1981 A neodymium and strontium isotopic study of the Mesozoic calc-alkaline granite batholiths of the Sierra Nevada, and Peninsular Ranges, California. *J. geophys. Res.* **86**, 10,470–10,488.

Dickinson, W. R. 1972 Evidence for plate-tectonic regimes in the rock record. *Am. J. Sci.* **272**, 551–576.

Dickinson, W. R. 1976 Sedimentary basins developed during evolution of Mesozoic-Cenozoic arc-trench systems in western North America. *Can. J. Earth Sci.* **13**, 1268–1289.

Dickinson, W. R., Ingersoll, R. V., Cowan, D. S., Helmold, K. P. & Suczek, C. A. 1982 Provenance of Franciscan graywackes in coastal California. *Geol. Soc. Am. Bull.* **93**, 95–107.

Dickinson, W. R., Ojakangas, R. W. & Stewart, R. J. 1969 Burial metamorphism of the Late Mesozoic Great Valley sequence, Cache Creek, California. *Geol. Soc. Am. Bull.* **80**, 519–526.

Donato, M. M. 1987 Evolution of an ophiolitic tectonic melange, Marble Mountains, northern California Klamath Mountains. *Bull. geol. Soc. Am.* **98**, 448–464.

Dumitru, T. A. 1989 Constraints on uplift in the Franciscan subduction complex from apatite fission track analysis. *Tectonics* **8**, 197–220.

Engebretson, D. C., Cox, A. & Gordon, R. G. 1985 Relative motions between oceanic and continental plates in the Pacific Basin. *Geol. Soc. Am. Spec. Paper* **206**.

Ernst, W. G. 1970 Tectonic contact between the Franciscan melange and the Great Valley sequence, crustal expression of a Late Mesozoic Benioff zone. *J. geophys. Res.* **75**, 886–901.

Ernst, W. G. 1976 *Petrologic phase equilibria.* San Francisco: W. H. Freeman.

Ernst, W. G. 1983 Phanerozoic continental accretion and the metamorphic evolution of northern and central California. *Tectonophys.* **100**, 287–320.

Ernst, W. G. 1984 Californian blueschists, subduction, and the significance of tectonostratigraphic terranes. *Geol.* **12**, 436–440.

METAMORPHISM IN TERRANES OF THE WESTERN U.S.

Ernst, W. G. 1987 Mafic meta-igneous arc rocks of apparent komatiitic affinities, Sawyers Bar area, central Klamath Mountains, northern California. In *Magmatic processes: physico-chemical principles* (ed. B. O. Mysen), pp. 191–208. Geochem. Soc. Spec. Pub. no. 1.

Ernst, W. G. (ed.) 1988 *Metamorphism and crustal evolution of the western United States.* Englewood Cliffs, New Jersey: Prentice-Hall.

Ernst, W. G., Seki, Y., Onuki, H. & Gilbert, M. C. 1970 Comparative study of low-grade metamorphism in the California Coast Ranges and the Outer Metamorphic Belt of Japan. *Geol. Soc. Am. Mem.* **124**.

Evarts, R. C. 1977 The geology and petrology of the Del Puerto ophiolite, Diablo Range, central California Coast Ranges. In *North American ophiolites* (ed. R. G. Coleman & W. P. Irwin), pp. 121–139. State of Oregon, Dept. Geol. Min. Indust., Bull. **95**.

Farmer, G. L. 1988 Isotope geochemistry of Mesozoic and Tertiary igneous rocks in the western U.S. and implications for the structure and composition of the deep continental lithosphere. In *Metamorphism and crustal evolution of the western United States* (ed. W. G. Ernst), pp. 87–109. Englewood Cliffs, New Jersey: Prentice-Hall.

Farmer, G. L. & DePaolo, D. J. 1983 Origin of Mesozoic and Tertiary granite in the western United States and implications for pre-Mesozoic crustal structure. 1. Nd and Sr isotopic studies in the geocline of the northern Great Basin. *J. geophys. Res.* **88**, 3379–3401.

Gray, G. G. 1986 Native terranes of the central Klamath Mountains, California. *Tectonics* **5**, 1043–1054.

Hackel, O. 1966 Summary of the geology of the Great Valley. In *Geology of northern California* (ed. E. H. Bailey), pp. 217–238. California Division of Mines and Geology, Bull. **190**.

Hamilton, W. 1969 Mesozoic California and the underflow of the Pacific mantle. *Geol. Soc. Am. Bull.* **80**, 2409–2430.

Hannah, J. L. & Moores, E. M. 1986 Age relationship and depositional environments of Paleozoic strata, northern Sierra Nevada, California. *Geol. Soc. Am. Bull.* **97**, 787–797.

Hill, M. L. & Dibblee, T. W. 1953 San Andreas, Garlock, and Big Pine faults, California. *Geol. Soc. Am. Bull.* **64**, 443–458.

Hopson, C. A., Mattinson, J. M. & Pessagno, E. A. Jr 1981 Coast Range ophiolite, western California. In *The geotectonic development of California* (ed. W. G. Ernst), pp. 418–510. Englewood Cliffs, New Jersey: Prentice-Hall.

Hsü, K. J. 1968 Principles of melanges and their bearing on the Franciscan–Knoxville paradox. *Geol. Soc. of Am. Bull.* **79**, 1063–1074.

Hsü, K. J. 1974 Melanges and their distinction from olistostromes. In *Modern and ancient geosynclinal sedimentation* (ed. R. H. Dott Jr & R. H. Shaver), pp. 321–333. *Soc. Econ. Paleont. Min., Spec. Publ.* **19**.

Ingersoll, R. V. 1979 Evolution of the Late Cretaceous forearc basin, northern and central California. *Geol. Soc. Am. Bull.* **90**, 813–826.

Irwin, W. P. 1972 Terranes of the western Paleozoic and Triassic belt in the southern Klamath Mountains, California. *U.S. Geol. Surv. Prof. Paper* **800-C**, 103–111.

Irwin, W. P. 1981 Tectonic accretion of the Klamath Mountains. In *The geotectonic development of California* (ed. W. G. Ernst), pp. 29–49. Englewood Cliffs, New Jersey: Prentice-Hall.

Irwin, W. P., Wolfe, E. W., Black, M. C. Jr & Cunningham, G. C. 1974 Geologic map of the Pickett Peak Quadrangle, Trinity County, California. *U.S. Geol. Surv. Geol. Quad. Map GQ-1111* (scale 1:62500).

Jacobson, M. I. 1978 Petrologic variations in Franciscan sandstone from the Diablo Range, California. In *Mesozoic paleogeography of the western United States* (ed. D. G. Howell & K. A. McDougall), pp. 401–417. Pacific Section, Soc. Econ. Paleont. Mineral., Pacific Coast Paleogeography Symp. **2**.

Jayko, A. S. & Blake, M. C. Jr 1984 Sedimentary petrology of graywacke of the Franciscan Complex in the northern San Francisco Bay area, California. In *Franciscan geology of northern California* (ed. M. C. Blake Jr), pp. 121–134. Pacific Section, Soc. Econ. Paleont. Mineral., **43**.

Jayko, A. S., Blake, M. C. & Harms, T. 1987 Attenuation of the Coast Range ophiolite by extensional faulting, and nature of the Coast Range 'thrust', California. *Tectonics* **6**, 475–488.

Jennings, C. W. 1977 Geologic Map of California. *Calif. Div. Mines Geol.* (scale 1:750000).

Jones, D. L., Howell, D. G., Coney, P. J. & Monger, J. W. H. 1983 Recognition, character, and analysis of tectonostratigraphic terranes in western North America. In *Accretion tectonics in the circumpacific region* (ed. M. Hashimoto & S. Uyeda), pp. 21–35. Tokyo: Terra Science Publishing.

Kistler, R. W. & Peterman, Z. E. 1978 A study of regional variations of initial strontium isotopic composition of Mesozoic granitic rocks in California. *U.S. Geol. Surv. Prof. Paper* **1071**.

Lanphere, M. A., Blake, M. C. Jr & Irwin, W. P. 1978 Early Cretaceous metamorphic age of the South Fork Mountain schist in the northern Coast Ranges of California. *Am. J. Sci.* **278**, 798–815.

Liou, J. G., Maruyama, S. & Cho, M. 1985 Phase equilibria and mineral parageneses of metabasites in low-grade metamorphism. *Mineral. Mag.* **49**, 321–333.

Luyendyk, B. P. 1982 Paleolatitude of the Point Sal ophiolite. *Geol. Soc. Am.* (*Abstr. with Programs*) **14**, 182.

Maruyama, S., Liou, J. G. & Sasakura, Y. 1985 Low-temperature recrystallization of Franciscan graywackes from Pacheco Pass, California. *Mineral. Mag.* **49**, 345–355.

111

Mattinson, J. M. 1988 Constraints on the timing of Franciscan metamorphism: geochronological approaches and their limitations. In *Metamorphism and crustal evolution of the western United States* (ed. W. G. Ernst), pp. 1023–1034. Englewood Cliffs, New Jersey: Prentice-Hall.

Mattinson, J. M. & Echeverria, L. M. 1980 Ortigalita Peak gabbro, Franciscan complex: U–Pb dates of intrusion and high-pressure-low-temperature metamorphism. *Geology* **8**, 589–593.

Maxwell, J. C. 1974 Anatomy of an orogen. *Geol. Soc. Am. Bull.* **85**, 1195–1204.

McKee, B. 1962*a* Widespread occurrence of jadeite, lawsonite, and glaucophane in central California. *Am. J. Sci.* **260**, 596–610.

McKee, B. 1962*b* Aragonite in the Franciscan rocks on the Pacheco Pass area, California. *Am. Mineral.* **47**, 379–387.

McLaughlin, R. J., Kling, S. A., Poore, R. Z., McDougall, K. & Beutner, E. C. 1982 Post-middle Miocene accretion of Franciscan rocks, northwestern California. *Geol. Soc. Am. Bull.* **93**, 595–605.

Misch, P. 1966 Tectonic evolution of the Northern Cascades of Washington State. In *Tectonic history and mineral deposits of the western Cordillera* (ed. H. C. Gunning), pp. 108–148. Canadian Inst. Mining Metallurgy Spec. V. **8**.

Miyashiro, A. 1973 *Metamorphism and metamorphic belts.* London: Allen & Unwin.

Oxburgh, E. R. & Turcotte, D. L. 1971 Origin of paired metamorphic belts and crustal dilation in island arc regions. *J. geophys. Res.* **76**, 1315–1327.

Page, B. M. 1982 Migration of Salinian composite block, California, and disappearance of fragments. *Am. J. Sci.* **282**, 1694–1734.

Peacock, S. M. 1988 Inverted metamorphic gradients in the westernmost Cordillera. In *Metamorphism and crustal evolution of the western United States* (ed. W. G. Ernst), pp. 953–975. Englewood Cliffs, New Jersey: Prentice-Hall.

Pitman, W. C., Larson, R. L. & Herron, E. M. 1974 The age of the ocean basins. *Geol. Soc. Am. Map Ser.* (horizontal scale 1:40000000).

Platt, J. P. 1986 Dynamics of orogenic wedges and the uplift of high-pressure metamorphic rocks. *Geol. Soc. Am. Bull.* **97**, 1037–1053.

Saleeby, J. B. 1981 Ocean floor accretion and volcanoplutonic arc evolution of the Mesozoic Sierra Nevada. In *The geotectonic development of California* (ed. W. G. Ernst), pp. 132–181. Englewood Cliffs, New Jersey: Prentice-Hall.

Sams, D. B. & Saleeby, J. B. 1988 Geology and petrotectonic significance of crystalline rocks of the southernmost Sierra Nevada, California. In *Metamorphism and crustal evolution of the western United States* (ed. W. G. Ernst), pp. 865–893. Englewood Cliffs, New Jersey: Prentice-Hall.

Schweickert, R. A. 1981 Tectonic evolution of the Sierra Nevada Range. In *The geotectonic development of California* (ed. W. G. Ernst), pp. 87–131. Englewood Cliffs, New Jersey: Prentice-Hall.

Schweickert, R. A., Merquerian, C. & Bogen, N. L. 1988 Deformational and metamorphic history of Paleozoic and Mesozoic basement terranes in the western Sierra Nevada metamorphic belt. In *Metamorphism and crustal evolution of the western United States* (ed. W. G. Ernst), pp. 789–822. Englewood Cliffs, New Jersey: Prentice-Hall.

Seiders, V. M. 1988 Origin of conglomerate stratigraphy in the Franciscan assemblage and Great Valley sequence, northern California. *Geology* **16**, 783–787.

Seiders, V. M. & Blome, C. D. 1984 Clast compositions of upper Mesozoic conglomerates of the California Coast Ranges and their tectonic significance. In *Franciscan geology of northern California* (ed. M. C. Blake Jr), pp. 135–148. Pacific Section, Soc. Econ. Paleont. Mineral., **43**.

Sharp, W. D. 1988 Pre-Cretaceous crustal evolution in the Sierra Nevada region, California. In *Metamorphism and crustal evolution of the western United States* (ed. W. G. Ernst), pp. 823–864. Englewood Cliffs, New Jersey: Prentice-Hall.

Shervais, J. W. & Kimbrough, D. L. 1985 Geochemical evidence for the tectonic setting of the Coast Range ophiolite: A composite island arc-oceanic crust terrane in western California. *Geology* **13**, 35–38.

Silberling, N. J., Jones, D. L., Blake, M. C. Jr & Howell, D. G. 1987 Lithotectonic terrane map of the western conterminous United States. *Miscellaneous Field Studies Map MF-1874-C* (scale 1:250000).

Smith, G. W., Howell, D. & Ingersoll, R. V. 1979 Late Cretaceous trench-slope basins of central California. *Geology* **7**, 303–306.

Suppe, J. 1972 Interrelationships of high-pressure metamorphism, deformation, and sedimentation in Franciscan tectonics, U.S.A. *Rep. 24th Int. Geology Congr., Montreal, Sec. 3*, pp. 552–559.

Suppe, J. 1973 Geology of the Leech Lake Mountain-Ball Mountain region, California. *Univ. Calif. Pub., Geol. Sci.* **107**, 1–82.

Suppe, J. & Armstrong, R. L. 1972 Potassium-argon dating of Franciscan metamorphic rocks. *Am. J. Sci.* **272**, 217–233.

Telleen, K. E. 1977 Paleocurrents in part of the Franciscan Complex, California. *Geology* **5**, 49–51.

Wentworth, D. M., Blake, M. C. Jr, Jones, D. L., Walter, A. W. & Zoback, M. D. 1984 Tectonic wedging associated with emplacement of the Franciscan assemblage, California Coast Ranges. In *Franciscan geology of northern California* (ed. M. C. Blake Jr), pp. 163–173. Pacific Section, Soc. Econ. Paleont. Mineral., **43**.

Wooden, J. L., Stacey, J. S., Doe, B. R., Howard, K. A. & Miller, D. M. 1988 Pb isotopic evidence for the formation of Proterozoic crust in the southwestern United States. In *Metamorphism and crustal evolution of the western United States* (ed. W. G. Ernst), pp. 69–86. Englewood Cliffs, New Jersey: Prentice-Hall.

Worrall, D. M. 1981 Imbricate low angle faulting in uppermost Franciscan rocks, South Yolla Bolly area, northern California. *Geol. Soc. Am. Bull.* **92**, 703–709.

Wright, J. E. 1982 Permo-Triassic accretionary subduction complex, southwestern Klamath Mountains, northern California. *J. geophys. Res.* **87**, 3805–3818.

Zartman, R. E. 1974 Lead isotope provinces in the Cordillera of the western United States and their geologic significance. *Econ. Geology* **69**, 792–805.

Zen, E-an 1988 Evidence for accreted terranes and the effect of metamorphism. *Am. J. Sci.* A **288**, 1–15.

Discussion

I. G. Gass, F.R.S. (*Open University, U.K.*). Professor Ernst claims the 'Yellow Dog' terrane came in (docked) 'steaming' and he describes dykes which invaded the terrane after it had accreted to western N. America. This implies that the terrane brought with it its own magma supply and perhaps heat source. In which case, what was the depth of cut off when the terrane became detached and does he regard such terranes as common or rare phenomenon? Is it possible that the magmatism was genetically associated with the N. American host? On what evidence did he decide that it was the allochthonous terrane that was steaming?

W. G. Ernst. The Sawyers Bar terrane consists of metamorphosed distal turbidites and interlayered – but predominantly overlying – Yellow Dog greenstones. Hypabyssal dikes and sills in the Yellow Dog metavolcanics are chemically and mineralogically identical, hence comagmatic with the extrusive rocks; the diabases are porphyritic and carry abundant relict phenocrysts of oscillatorily zoned hornblende. These same distinctive hypabyssals, on a lesser scale, transect chemically, mineralogically different rocks of the more easterly, structurally higher Stuart Fork late Triassic blueschist complex. Accordingly, the outboard (westerly) Sawyers Bar immature island-arc terrane must have been still active (steaming) as it collided with the previously deformed and metamorphosed, landward Stuart Fork complex, presumably along the western margin of North America. Because early–middle Jurassic regional greenschist facies metamorphism of both terranes accompanied this suturing event, and because thermobarometric data suggest physical conditions on the order of 400 °C at about 3 kilobar (3×10^8 Pa), the depth of decoupling of the Sawyers Bar section from the subducting Palaeopacific plate must have been approximately 10 km. I suspect that such events are common in the Circumpacific area in cases where marginal basin collapse has allowed the docking of offshore arcs along adjacent continental margins.

A. H. F. Robertson (*Grant Institute of Geology, University of Edinburgh, U.K.*). From my own field studies, I would strongly support Professor Ernst's view that the Great Valley Sequence and the Franciscan were essentially coupled as in classical plate tectonic interpretations; in particular, there is no field structural evidence of significant strike–slip faulting near the contacts. However, how would he envisage the origin to the ophiolitic slivers and eclogitic rocks within the Franciscan; were they slivered off the edge of the Coast Range ophiolite and transported northwards by strike–slip faulting, as has been suggested, or are there other explanations?

W. G. ERNST. According to many authors (e.g. Shervais & Kimbrough 1985), the bulk-rock geochemistry of basaltic/gabbroic/serpentinitic blocks contained within the Franciscan contrasts with that of the Coast Range ophiolite lying beneath the Great Valley Sequence. Thus, although the latter may represent the westward, more oceanic part of the North American lithospheric plate, Franciscan meta-ophiolite tectonic blocks probably were derived from a Palaeopacific substrate on which the voluminous terrigenous debris was deposited, then disaggregated in the subduction zone. Alternatively, such tectonic lenses may represent hanging-wall lithologies metamorphosed under high-P (and moderately high-T) conditions at the initiation of subduction in late(?) Jurassic time.

Allochthonous terranes of the Southwest Pacific and Indonesia

By M. G. Audley-Charles[1] and R. A. Harris[2]

[1] Department of Geological Sciences, University College London, Gower Street, London WC1E 6BT, U.K.

[2] Department of Geology and Geography, University of West Virginia, Morgantown, West Virginia 26506, U.S.A.

Rift–drift processes associated with the Mesozoic break-up of Gondwana and subsequent collisional events involving rifted crustal blocks from Gondwana in Tethys led to the formation and emplacement of allochthonous terranes in fold and thrust mountain belts. These terranes include small allochthonous continental blocks (*ca.* 1000 km²) and allochthonous exotic blocks that may exceed 200 km². Other processes consequent upon the break-up of Gondwana and subsequent plate convergence between Gondwana and Asia include the emplacement of supra-subduction zone ophiolites as allochthonous terranes superimposed on the continental margin. The accretion of oceanic plateaus to continental and arc terranes in the Western Pacific have resulted from trenches and strike–slip faults being deflected by crustal heterogeneities. Indonesian and Western Pacific regions display allochthonous terranes forming and being emplaced at rates similar to plate movements. Many allochthonous terranes of the SW Pacific and eastern Indonesia have been accreted during the past 3 Ma and, being so young, have not suffered overprinting by later tectonic and thermal events. Furthermore, we can trace some of the very young nappes into their roots and trace some accretionary processes into zones where, at present plate convergence rates, the collision and hence accretion will not occur for about another 1 Ma. These active regions reveal the importance of local events affecting less than about 1000 km of plate boundary over a period of 1–2 Ma.

INTRODUCTION

In their world-wide review of allochthonous terranes and their role in the evolution of fold and thrust mountain belts Nur & Ben-Avraham (1982) identified oceanic plateaus in all the main oceans, and continental fragments, mainly in and around the Pacific and Indian oceans, as likely candidates for conversion into allochthonous terranes. Coney *et al.* (1980) identified three main crustal types of allochthonous terranes already emplaced in the cordillera of western North America: oceanic, oceanic volcanic arc, and parts of distal continental edges. Some parts of the Wrangellia terrane may have been oceanic plateaus and Cache Creek terrane may include carbonate atolls built on seamounts. Coney *et al.* (1980) pointed out that some existing allochthonous terranes revealed clear indications of having undergone tectonic amalgamation before emplacement and accretion to the North American cordillera. This is also apparent among the SE Asian terranes. However, one notable feature of the terranes accreted to the NE Australian margin is that many of them show strong continental margin affinities in contrast to the dominantly oceanic origin of the Western Cordillera terranes.

The SW Pacific and Indonesian margins of Australia (figure 1) have undergone phases of major continental rifting with both large-scale (10³ km) and much smaller-scale (10² km)

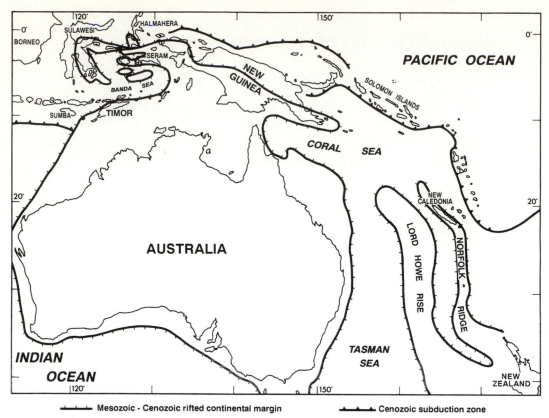

FIGURE 1. Eastern and northern margins of Australia showing the Mesozoic–Cenozoic rifted continental margins of Australian Gondwana. Note the continental blocks removed by Tasman Sea spreading. Note the post-rifting collisions with allochthonous terrane emplacement in eastern Indonesia, New Guinea and New Caledonia shown in figures 2 and 3.

spreading during the late Mesozoic and early Cenozoic (Falvey & Mutter 1981). The SW Pacific region has been characterized during the late Cretaceous and Cenozoic by plate convergence associated with subduction and volcanism. These destructive plate margin processes have influenced the northern Australian margin only since the end of the Palaeogene, related to Australia's 3000 km northward drift. The eastern and northern margins of Australia–New Guinea (figures 2 and 3) provide examples of the active processes of allochthonous terrane formation of all the three main crustal types identified by Coney *et al.* (1980) from the cordillera of western North America. It is possible to recognize all three types for example in the 3 Ma old fold and thrust belt of the Outer Banda Arc and similar allochthonous terranes have been reported from elsewhere along the collisional boundaries of Tethys. There are other types of allochthonous terrane in eastern Indonesia such as the exotic block terranes of Permo–Triassic shallow marine limestones (Maubisse facies) and nappes of volcanic arc accreted to continental crust (Sumba–Palelo arc); both are described below.

Large continental allochthonous terranes

By repeated rifting the eastern part of the supercontinent Gondwana broke up into large continental-sized blocks such as India, Australia and Indochina and relatively thin, long and narrow continental slivers such as Sibumasu, which includes parts of South Tibet, Burma,

116

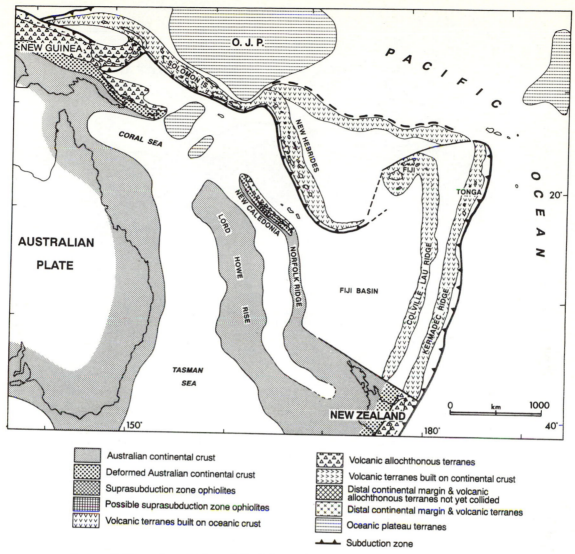

FIGURE 2. Tectonic accretion of allochthonous terranes to rifted continental margins
of eastern Australia and New Guinea.

western Thailand, Malay Peninsula and Sumatra (Metcalfe 1988), as well as other long, narrow slivers (figure 1) but with thinner continental crust (Shor *et al.* 1971) represented by the Lord Howe Rise, the New Caledonia–Norfolk Ridge (Kronke 1984), both of which are largely submarine. All these rifted continental blocks and slivers have surface dimensions measuring thousands of kilometres long and several hundreds of kilometres wide. These allochthonous terranes are comparable in size with some Canadian and Alaskan 'cordilleran suspect terranes' of western North America (Coney *et al.* 1980; Nur & Ben-Avraham, 1982). The processes of formation and subsequent accretion of some of these larger Gondwana continental blocks and slivers into the continental collage of Asia (Metcalfe 1988; Audley-Charles *et al.* 1988) provide analogies with the detachment of much smaller continental blocks and slivers having surface dimensions measured in tens of kilometres (figure 4), which were accreted during the late Cenozoic into the evolving fold and thrust belts at colliding continental margins in eastern Indonesia.

117

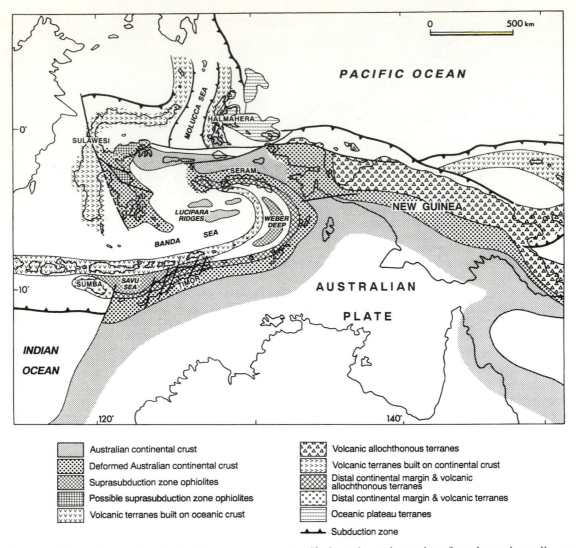

FIGURE 3. Tectonic accretion of allochthonous terranes to rifted continental margins of northeast Australia, and oceanic plateau–arc collisions.

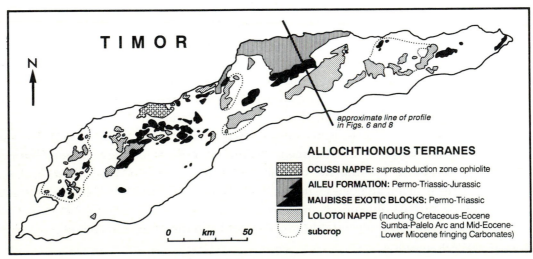

FIGURE 4. Allochthonous terranes in Timor.

118

The Outer Banda Arc fold and thrust belt (figure 2) is composed of two main elements: the deformed margin of the Australian continent and an overriding series of allochthonous nappes (Price & Audley-Charles 1987). Within this mountain belt in Timor the lowest nappes (figure 4) have a metamorphic basement about 2 km thick overlain by a Cretaceous–Eocene volcanic and sedimentary cover about 1–2 km thick, overlain unconformably mainly by shallow marine limestones ranging from middle Eocene to early Miocene. Harris (1989) has shown that the sedimentary and volcaniclastic Lower Cretaceous Palelo sequence grades down into the monometamorphic sequence of greenschists to amphibolites forming the upper part of the Lolotoi, below which are the polyphase gneisses (Earle 1981). Locally, serpentinite and tectonized peridotite form the top of the Lolotoi Complex. In Seram, the other large island of the Outer Banda Arc, similar nappes occur having a metamorphic basement of Kaibobo and Kobipoto Complex (Audley-Charles et al. 1979).

The allochthonous elements form a forearc basement that is found thrust over the deformed Australian continental margin sequence in Timor (Price & Audley-Charles 1987; Harris, 1989) composed of Permian to early Pliocene cover rocks. These allochthonous volcanic and volcaniclastic rocks may be traced, via a forearc submarine ridge from northwest Timor westwards, into Sumba where these rocks are now exposed as part of the forearc basement of the Banda volcanic arc. New exposures of the overthrust contact in Timor have been revealed in a road cutting through the Lolotoi nappe north of the village of Same. Audley-Charles (1985) demonstrated a close correlation of the Cretaceous to early Miocene deposits of the island of Sumba (figure 4) with the cover rock sequence of the Lolotoi in Timor (and by implication in Seram).

This led Harris (1989) to postulate that the forearc basement of the present Banda volcanic arc was formerly part of a Sumba–Palelo Arc, in which the late Cretaceous–Eocene volcanic forearc deposits accreted to the Lolotoi Complex metamorphic basement (figures 5 and 6). Both Hall (1988) and Harris (1989), noting the mineralogical evidence (Brown & Earle 1983) for rapid uplift of the Lolotoi, which is characterized by (HT/LP) metamorphism, have suggested that the high-temperature metamorphism of the Lolotoi Complex can be most easily explained by the thermal events of the late Jurassic–early Cretaceous rifting and spreading at the Australian continental margin. This is supported by the early Cretaceous isochron age for the polyphase gneiss (data reviewed by Harris (1989)). Thus the Lolotoi allochthonous terrane is viewed as having originated as a narrow sliver from the distal continental margin of Australian Gondwana rifted and separated by Tethyan spreading in the late Jurassic–early Cretaceous (figure 6). The K–Ar cooling ages from the Lolotoi suggested that the Lolotoi was accreted to the Sumba–Palelo volcanic arc in the late Oligocene, followed by the uplift associated with deposition of the unconformably overlying Oliogo–Miocene Cablac Limestone.

Later this Lolotoi terrane, as part of the Banda allochthon, was thrust over the Australian continental margin of Australia in Timor, Seram and some smaller islands during the latest Miocene (Carter et al. 1976) to the present as the collision progressively involves the NW Australian continental margin. The nappe emplacement was a consequence of the collision between the Australian margin and the Banda volcanic arc (figure 7) which, from 20 Ma to the present, has been built on the backarc of the defunct Sumba–Palelo arc (Harris 1989). These Lolotoi terranes (figures 4 and 8) are composed of small fragments (having surface

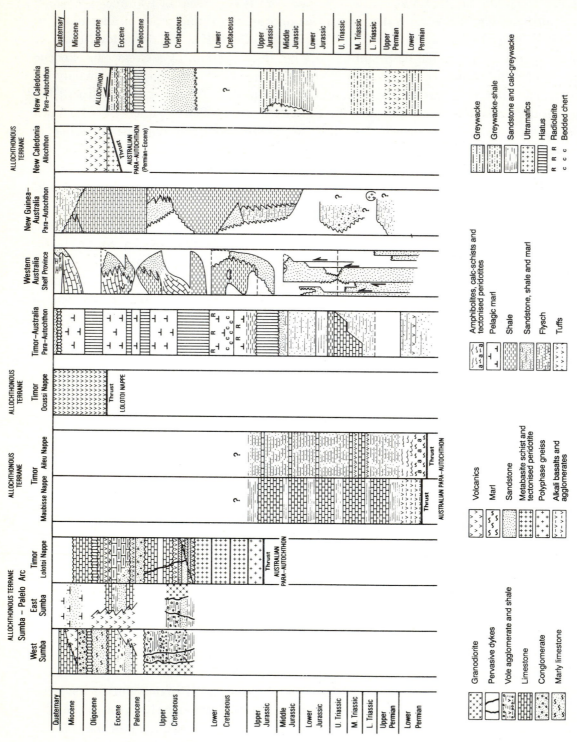

FIGURE 5. Stratigraphic summary of allochthonous terranes in the Timor and New Caledonia, comparison with Sumba and contrasts with the Australian para-autochthon in Timor, Australian Shelf, New Guinea and New Caledonia. Data summarized as follows: Sumba from Audley-Charles (1985); Timor from Audley-Charles (1978, 1985, 1988); Western Australia and New Guinea from Falvey & Mutter (1981), New Caledonia from Guillon (1974).

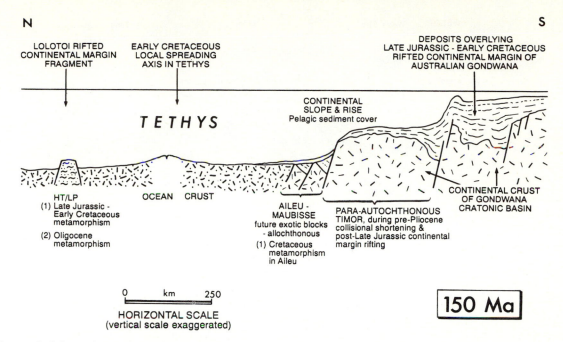

FIGURE 6. Schematic representation of the formation of the Lolotoi and Aileu–Maubisse allochthonous terranes at the Australian continental margin. The Sumba–Palelo volcanic arc was built on the Lolotoi continental fragment during the Cretaceous–Eocene after it had separated from the Gondwana continental margin. See figure 4 for the approximate position of this profile.

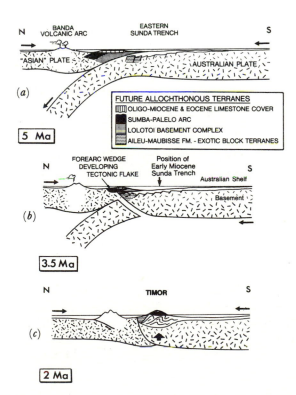

FIGURE 7 a, b, c. Schematic representation (after Price & Audley-Charles 1987) of accretion of allochthonous terranes to the Australian margin in the Banda Arc.

121

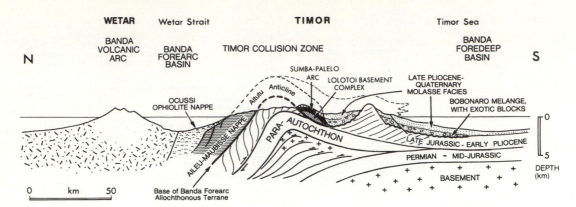

FIGURE 8. Section through Timor and collided Banda volcanic arc to show present position of accreted allochthonous terranes (after Harris 1989). See figure 4 for the position of this profile.

dimensions of a few tens of kilometres by a few hundreds of kilometres of the Gondwana continental margin on which has been preserved part of the Cretaceous–Palaeogene volcanic forearc and later fringing shallow marine limestones. These terranes have been much attenuated by numerous low-angle normal faults (Harris 1989) and also much broken by steep normal faults during the Quaternary.

Similar collisional phenomena have been reported from New Guinea by Pigram & Davies (1988) who have identified more than 10 small continental allochthonous terranes accreted to the Australian margin during the Oligo–Miocene.

EXOTIC BLOCKS ALLOCHTHONOUS TERRANES

One of the features of the Tethyan mountain belt is the presence of exotic blocks. One of the most characteristic type of blocks is the highly fossiliferous Permo–Triassic shallow marine limestones associated with pillow lavas, for example in the Himalayas (Marcoux *et al.* 1982; Searle 1983) and Oman (Glenie *et al.* 1974; Searle & Graham, 1982). These blocks are exotic in being allochthonous, in having no known roots and no easily detectable provenance, although Lippard *et al.* (1986) have proposed seamount carbonate cappings for the Oman occurrences. In the Banda Arc these exotic blocks are closely associated with the Bobonaro coloured scaly clay melange in which they appear to be engulfed, and with which they seem to have been emplaced above the Australian continental margin deposits (Audley-Charles 1986). The exotic blocks in Timor (e.g. Mt Lacouse and Mt Legumau) have surface dimensions that in some cases exceed 200 km² (figure 4).

In the Tethyan fold belt, and certainly in the Banda Arc these exotic blocks surrounded by coloured scaly clay melange occupy the highest structural position of the allochthonous elements. In the Banda Arc the most common exotic blocks belong to the Maubisse Formation that accumulated in an Australian Gondwana cratonic basin during the Permo–Trias (Bird 1987; Audley-Charles 1988). The mechanism by which they became detached from the Gondwana margin is uncertain, but it must have been a very widespread phenomenon of the Gondwana–Tethyan margin associated with the last episode of continental rifting because the exotic blocks were preserved near the foot of the continental slope. From there they were

122

carried back by the tectonic collision over the Gondwana margin in a melange along much of its length from the Mediterranean orogens via Oman and Himalayas to Timor and Seram. We suggest they became detached from the Gondwana continental margin, either by large-scale oceanwards slumping by processes analogous to the oceanward displacement of large slump blocks along parts of the Atlantic margins (Dingle 1977), or by the rifting leaving them as the most distal parts of the continental margin.

In the Banda Arc these Permo–Triassic limestone blocks associated with shales and vesicular alkaline basalts (Maubisse facies) appear to have formed the most distal part of the newly rifted continental slope during the late Jurassic–early Cretaceous. Here they were engulfed by deep-sea pelagic sediments of the slope and rise during the late Mesozoic and much of the Cenozoic. Locally, the stratigraphical passage from these shallow marine limestones into a flysch facies (Aileu Formation) of probably Permian to Jurassic age (Brunnschweiler 1978; Barber *et al.* 1977) has been preserved in central Timor. Amphibolites in this Aileu Formation indicate an Ar–Ar metamorphic event more than 70 Ma ago (Berry & McDougall 1986). This would correspond with the continental rifting thermal event comparable with that in the Lolotoi described above. In late Miocene–early Pliocene times the Australian continental margin collided with the Banda volcanic forearc. Parts of these rifted blocks together with part of the engulfing Mesozoic–Cenozoic pelagic sediment (Margolis *et al.* 1978), lying at the foot of the Australian continental rise, were carried over the lowest nappes by the converging Asian volcanic arc forearc ramping up onto the contracting Australian continental slope and rise.

SSZ OPHIOLITE ALLOCHTHONOUS TERRANES

Another kind of allochthonous terrane process that characterizes the closure of Tethys and which has been active in the SW Pacific and Indonesia during the last 3 Ma is the emplacement of the suprasubduction zone ophiolites. In eastern Indonesia it is possible to trace the Ocussi ophiolite nappe, now being thrust over the passive continental margin in northern Timor, back into its root zone in the Wetar Strait which is the forearc basin of the Banda volcanic arc. This Banda forearc basin is related to the eastern Sunda Trench with which the Australian continental margin collided at 3 Ma. One effect of this collision was the forearc ramping up and over (figure 7a, b) the converging continental margin thus closing the trench along the collision suture (Price & Audley-Charles 1987).

Harris (1989) has shown that the clinopyroxene–phyric basalts and andesite-basalts of the Ocussi nappe and the samples dredged from the floor of the Wetar Strait are part of a low-K tholeiite series having trace element affinities with island arc tholeiites. He argued from their geochemical signature and from palaeogeographical considerations that they were most likely formed as new spreading crust in the upper plate adjacent to irregularities in the shape of the lower plate in the collision zone. Transtensional forces associated with the regions sandwiched between the collisional indenters of Seram and east Timor led to the formation of small rhombochasmic basins (Savu and Weber basins), which are now in the process of tectonic emplacement. The Ocussi nappe is the initial manifestation of the emplacement process of the Savu basin as an allochthonous terrane, which is still in contact with its roots in the forearc basin. The formation and predicted emplacement of these basins is most likely a function of continentward trench retreat into embayed regions of the lower plate where young, thin parts of the forearc ramp up over the passive continental margin (figure 7). Although little is known about the crustal structure and age of these basins the data available (Harris 1989) suggests they are synorogenic and may represent some of the only modern analogues of Tethyan-type ophiolites.

ROLE OF STRIKE–SLIP FAULTS IN TERRANE FORMATION

Observations in the SW Pacific and Indonesian region suggests that there are at least two main types of strike–slip generated terrane. Plate boundary strike–slip faulting associated with oblique plate convergence has been suggested (Hamilton 1979) as a mechanism by which slivers of lithosphere become detached and are transported along the obliquely converging plate boundary. The Sorong Fault Zone appears to have removed crustal (or lithospheric) blocks from the northwest margin of New Guinea and to have transported them westwards (Visser & Hermes 1962). The evidence for identifying these terranes has been comparable stratigraphic sequences and characteristic magmatic products. Very little palaeomagnetic data exists against which to test some of these palaeogeograhic interpretations (Haile & Briden 1984; Metcalfe 1988). Examples of terranes thought to have been removed from New Guinea in this way include Banggai-Sula (Klompe 1954), eastern Sulawesi and Buton (Audley-Charles *et al.* 1972; Audley-Charles 1988), Obi (Silver & Smith 1983) and Bacan (Hall & Nichols 1990).

Another type of strike–slip fault mechanism influencing terrane distribution has been suggested by Charlton (1988) whereby faults cutting obliquely across the converging active plate margin off-set and so separate the converging allochthons. He applied this model to the southern Banda Arc collision zone, particularly in the Timor region.

ACCRETION OF OCEANIC PLATEAUS

One of the discoveries made by the early analyses of allochthonous terranes in the western Cordillera of North America was the relative importance of what appear to be remnants of oceanic plateaus. Studies in the Western Pacific (Hall & Nichols 1990) have revealed the ways in which the thickened crust of these oceanic plateaus can deflect the path of propagating trenches and strike–slip faults and lead to the relatively passive accretion of oceanic plateaus at evolving plate boundaries. Hall & Nichols (1990) have argued that East Mindanao, the Snellius Ridge and the East Halmahera–Waigeo terrane are oceanic plateaus at different stages of amalgamation into the Philippine margin. The complex amalgamation experienced by these terranes before their eventual accretion to the Asian or Gondwana continental margin affords comparison with some of the complex oceanic terranes described from the western Cordillera of North America (Coney *et al.* 1980).

The Ontong Java Plateau is one of the largest oceanic plateaus. It is now impinging on the eastern boundary of the Solomon island arc. It has been suggested (Hughes & Turner 1977) that the difficulty in subducting such a thick slab led to subduction polarity reversal below the Solomon arc in the Tertiary. Following the discoveries of Hall & Nichols (1990) in the Halmahera region (figure 9) continuing plate convergence between Ontong Java and the Solomons seems likely to lead to the development of a new active plate margin on the Pacific side of this huge oceanic plateau with the relatively passive accretion of the Ontong Java Plateau to the Arc. Pigram & Davies (1988) have identified many oceanic allochthonous

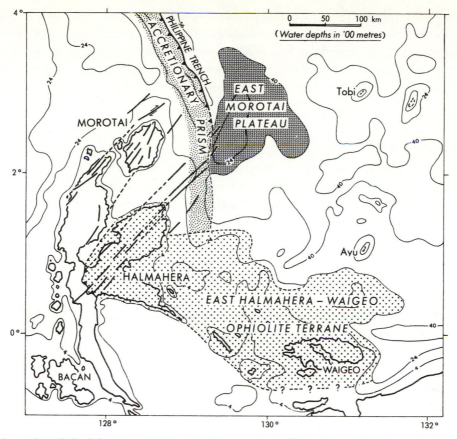

FIGURE 9. Accretion of allochthonous oceanic terranes to arcs in the Halmahera region (from Hall & Nichols 1990).

terranes derived from seamounts and oceanic plateaus that were accreted to New Guinea in the Oligo–Miocene.

RATES OF ALLOCHTHONOUS TERRANE EMPLACEMENT

Many Mesozoic–Palaeozoic fold and thrust mountain belts display indications of orogenesis involving the emplacement and movement of allochthonous terranes associated with collision and strike–slip deformation of the continental margin extending over several 10^7 years. Studies of these processes in the SW Pacific and Indonesia reveal that collisional orogens involving the evolution of allochthonous nappes moving *ca.* 50 km over deforming passive continental margins at *ca.* 8 cm a^{-1} can be created in less than 3 Ma. Furthermore, active plate margin reversal can occur within 2 Ma (Price & Audley-Charles 1987).

IMPORTANCE OF LOCAL TECTONIC EVENTS

One of the processes of active arc-trench systems of the Western Pacific and Indonesia is the relative importance of the local collisional event affecting less than 1000 km of plate boundary over a period of 1–2 Ma. Selected examples include New Caledonia, D'Entrecasteau and Louisville ridges, and the Timor, Seram and East Sulawesi sectors of the Banda Arc. This suggests that the regional scale of collision may be less extensive in some Mesozoic–Palaeozoic

125

orogens than is generally supposed. There are indications from the western Pacific, Philippines and Indonesia that one or several associated magmatic arcs generated in the Philippine Sea region may have been dispersed by intra-arc spreading and rotation (Haston *et al.* 1988; Lewis *et al.* 1982). Reconstruction of such palaeogeographical details after dispersed arcs have collided separately at different margins may be beyond resolution in ancient orogenic belts.

CONCLUSIONS

Allochthonous terranes can be observed in the SW Pacific, Philippine and eastern Indonesian regions in the process of formation. These active terranes closely resemble the main types described from the cordillera of western North America. However, an outstanding feature of the SW Pacific and Indonesian region is the high proportion of continental allochthonous terranes. This may be a consequence of the weakening of the continental lithosphere by repeated rifting of blocks and slivers from Gondwana. Study of these active regions suggests that the construction of a major accretionary cordilleran fold and thrust mountain system proceeds by many small local collisions (less than *ca.* 1000 km strike length). These active regions also indicate that the collision processes involving subduction zone reversal, or change in position of trench or major strike–slip fault related to terrane accretion can occur within 2 Ma. The apparent synchroneity and interpreted regional correlation of mountain building processes in ancient orogens is partly an artefact of the dating imprecision and partly the result of difficulties in resolving palaeogeographies from overprinted tectonic events in an eroded complex accretionary cordillera.

Janet Baker and Colin Stuart produced all the art work.

REFERENCES

Audley-Charles, M. G. 1978 The Indonesian and Philippine archipelagos. In *The Phanerozoic geology of the world II. The Mesozoic, A.* (ed. M. Moullade & A. E. M. Nairn), pp. 165–207. Amsterdam: Elsevier.

Audley-Charles, M. G. 1985 The Sumba enigma: is Sumba a diapiric forearc nappe in process of formation? *Tectonophysics* **119**, 435–449.

Audley-Charles, M. G. 1986 Rates of Neogene and Quaternary tectonic movements in the Banda arc based on micropalaeontology. *J. geol. Soc. Lond.* **143**, 161–175.

Audley-Charles, M. G. 1988 Evolution of the southern margin of Tethys (North Australian region) from early Permian to Late Cretaceous. In *Gondwana and Tethys* (ed. M. G. Audley-Charles & A. Hallam) *Geol. Soc. Lond. Spec. Rep.* **37**, 79–100.

Audley-Charles, M. G., Carter, D. J. & Milsom, J. S. 1972 Tectonic development of eastern Indonesia in relation to Gondwanaland dispersal. *Nature, Lond.* **240**, 137–139.

Audley-Charles, M. G., Ballantyne, P. D. & Hall, R. 1988 Mesozoic–Cenozoic rift–drift sequence of Asian fragments from Gondwanaland. *Tectonophysics* **155**, 317–330.

Barber, A. J., Audley-Charles, M. G. & Carter, D. J. 1977 Thrust mechanics in Timor. *J. geol. Soc. Australia* **24**, 51–62.

Berry, R. F. & McDougall, I. 1986 Interpretations of ^{40}Ar/^{39}Ar dating evidence from the Aileu Formation, East Timor, Indonesia. *Chem. Geol.* **59**, 43–58.

Bird, P. 1987 Stratigraphy, sedimentology and structure of the Kekneno region of west Timor and its relationship with the northwest shelf of Australia. Ph.D. thesis, University of London, U.K.

Brown, M. & Earle, M. M. 1983 Cordierite-bearing schists and gneisses from Timor, eastern Indonesia: *P–T* conditions of metamorphism and tectonic implications. *J. metamorphic Geol.* **1**, 183–203.

Brunnschweiler, R. O. 1978 Notes on the Geology of East Timor. *Bull. Bureau Mineral Resources* **192**, 9–1.

Carter, D. J., Audley-Charles, M. G. & Barber, D. J. 1976 Stratigraphical analysis of island arc-continental margin collision in eastern Indonesia. *J. geol. Soc. Lond.* **132**, 179–198.

Charlton, T. R. 1989 Stratigraphic correlation across an arc-continent collision zone: Timor and the Australian northwest shelf. *Austral. J. Earth Sci.* **36**, 263–274.

Coney, P. J., Jones, D. L. & Monger, J. W. H. 1980 Cordilleran suspect terranes. *Nature, Lond.* **288**, 329–333.

Dingle, R. V. 1977 The anatomy of a large submarine slump on a sheared continental margin (southeast Africa). *J. geol. Soc. Lond.* **3**, 293–310.

Earle, M. M. 1981 A study of Booi and Molo, two metamorphic massifs on Timor, eastern Indonesia. Ph.D. thesis, University of London, U.K.

Falvey, D. A. & Mutter, J. C. 1981 Regional plate tectonics and the evolution of the Australia's passive continental margins. *BMR J. Aust. Geol. Geophys.* **6**, 1–29.

Glennie, K. W., Bouef, M. G. A., Hughes-Clark, M. W., Moody-Stuart, M., Pilaar, W. F. J. & Reinhardt, B. M. 1974 Geology of the Oman Mountains, Verh. K. Ned. Geol-Mijnb. Genoot. (423 pages.)

Guillon, J. H. 1974 New Caledonia. In *Mesozoic–Cenozoic Orogenic Belts* (ed. A. M. Spencer) *Geol. Soc. Lond. Spec. Rep.*, **4**, 445–452.

Haile, N. S. & Briden, J. C. 1984 Past and future palaeomagnetic research and the tectonic history of east and southeast Asia. Palaeomagnetic research in southeast and East Asia. UN/ESCAP, *CCOP Tech. Bull. Publ.* **13**, 25–46. (Bangkok).

Hall, R. 1988 Plate boundary evolution in the Halmahera region, Indonesia. *Tectonophysics* **144**, 337–352.

Hall, R. & Nichols, G. J. 1990 Terrane accretion in the Philippine Sea margin. *Tectonophysics*. (In the press.)

Harris, R. A. 1989 Processes of allochthon emplacement, with special reference to the Brooks Range ophiolite, Alaska and Timor, Indonesia. Ph.D. thesis, University of London, U.K.

Haston, R., Fuller, M. & Schmidtke, E. 1988 Paleomagnetic results from Palau, west Caroline islands: a constraint on Philippine Sea plate motion. *Geology* **16**, 654–657.

Hughes, G. W. & Turner, C. C. 1977 Upraised Pacific Ocean floor, southern Malaita, Solomon islands. *Bull. geol. Soc. Am.* **88**, 412–424.

Klompe, T. H. F. 1954 The structural importance of the Sula Spur (Indonesia). *Indones. J. nat. Sci.* **110**, 21–40.

Kroenke, L. W. 1984 Cenozoic development of the southwest Pacific. UN/ESCAP, *CCOP/SOPAC Tech. Bull.* **6**, 1–122.

Lewis, S. D., Hayes, D. E. & Mrozowski, C. L. 1982 The origin of the West Philippine basin by inter-arc spreading. In *Geology and tectonics of the Luzon–Marianas region.* (ed. G. R. Balce & A. S. Zanoria). Philippines SEATAR Committee Spec. Publ. **1**, 31–51.

Lippard, S. J., Shelton, A. W. & Gass, I. G. 1986 The ophiolite of northern Oman. *Spec. Publ. Geol. Soc. Lond. Mem.* **11**. (178 pages.)

Marcoux, J., Mascle, G. & Cuif, J.-P. 1982 Existence de marqueurs bio-sedimentaires et structuraux tethysiens issus de la marge gondwanienne a la bordure ouest-americaine: implications paleogeographiques et paleo-biologiques. *Bull. Soc. géol. Fr.* **7**, 971–980.

Margolis, S. V., Ku, T. L., Glasby, G. P., Fein, C. D. & Audley-Charles, M. G. 1978 Fossil manganese nodules from Timor: geochemical and radiochemical evidence for deep-sea origin. *Chem. Geol.* **21**, 185–198.

Metcalfe, I. 1988 Origin and assembly of southeast Asian continental terranes. In *Gondwana and Tethys* (ed. M. G. Audley-Charles & A. Hallam). *Geol. Soc. Lond., Spec. Rep.* **37**, 101–118.

Nur, A. & Ben-Avraham, Z. 1982 Oceanic plateaus, the fragmentation of continents, and mountain building. *J. geophys. Res.* **87**, 3644–3661.

Pigram, C. J. & Davies, H. L. 1988 Terranes and the accretion history of the New Guinea orogen. *BMR J. Aust. Geol. Geophys.* **10**, 193–211.

Price, N. J. & Audley-Charles, M. G. 1987 Tectonic collision processes after plate rupture. *Tectonophysics* **140**, 121–129.

Searle, M. P. 1983 Stratigraphy, structure and evolution of the Tibetan-Tethys zone in Zanskar and the Indus suture zone in the Ladakh Himalaya. *Trans. R. Soc. Edin. Earth Sci.* **73**, 205–219.

Searle, M. P. & Graham, G. M. 1982 The 'Oman Exotics': oceanic carbonate build-ups associated with the early stages of continental rifting. *Geology* **10**, 43–49.

Shor, G. G. Jr, Kirk, H. K. & Menard, H. W. 1971 Crustal structure of the Melanesian area. *J. geophys. Res.* **76**, 2562–2586.

Silver, E. A. & Smith, R. B. 1983 Comparison of terrane accretion in modern southeast Asia and the Mesozoic North American Cordillera. *Geology* **11**, 198–202.

Visser, W. A. & Hermes, J. J. 1962 Geological results of the exploration for oil in Netherlands New Guinea. *K. Ned. Geol. Mijnb. Genoot. Verh. geol. ser.* **20**, 1–265.

Discussion

P. D. CLIFT (*Grant Institute, University of Edinburgh, U.K.*). The proposed model for the generation of ophiolites with a suprasubduction zone chemistry (ssz), in a back arc basin and being emplaced on to the continental margin as the Australian plate comes in to collision with the Banda Arc contrasts markedly with many well-known ophiolites from the Tethyan and Caledonide realms. The Bay of Islands, Semail, Turkish and Greek ophiolites, although partly comprising mafic extrusives whose trace element geochemistry is of ssz type, are not apparently associated with well-developed arcs. Any arcs involved have either been destroyed by subduction (although density contrasts appear to prohibit this) or were never formed in the first place. These ophiolites appear to have been generated and soled shortly after the start of intra-oceanic subduction and were then subsequently emplaced (figure 10). It is important to note that in almost no instance does ophiolite emplacement accompany continental or arc collision events, often predating these by more than 100 Ma.

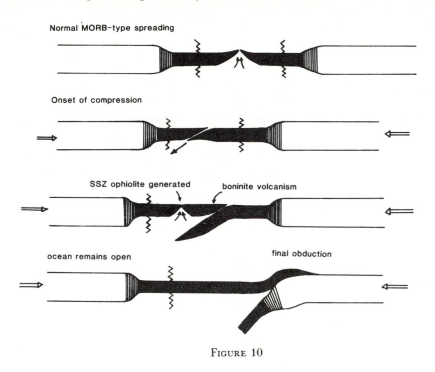

Normal MORB-type spreading

Onset of compression

SSZ ophiolite generated   boninite volcanism

ocean remains open   final obduction

FIGURE 10

M. G. AUDLEY-CHARLES. Dr Clift's question raises two separate issues: (*a*) the specific matter of the origin and emplacement of the ssz ophiolites in northern Timor, and (*b*) the more general case of the origin and emplacement of Tethyan-type ophiolites

(a) *Origin and emplacement of Timor Tethyan-type ophiolites having ssz geochemistry*

The top of what we suspect is a large ophiolite appears in northern Timor as the Ocussi nappe (Carter *et al.* 1976). These clinopyroxene–phyric basalt and basaltic–andesite pillows, sheet flows and breccias are part of a low-K tholeiite series having trace element affinities with island arc tholeiites (Harris 1989). Ar–Ar dating suggests these rocks are Oligo–Miocene in age and their stratigraphic cover indicates they must be no younger than upper Miocene. These rocks appear to floor the Banda forearc basin of the Wetar Strait from where samples have been dredged, and they must extend offshore northern Timor to floor part of the Savu Sea Banda forearc basin. Harris (1989) has interpreted the Ocussi nappe as the initial manifestation of emplacement of the Savu forearc basin. This basin is thought to have formed by the retreat of a subduction trench into embayments of an irregular shaped continental margin during collision. The uplift of the subducted continental margin emplaces what we interpret as the ssz ophiolite of the hanging wall forearc as a roof thrust (Price & Audley-Charles, 1983).

If a continental margin is old, as is usually the case with Tethyan-type ophiolites the attached oceanic slab, which has already been subducted, will define a steep Benioff zone as in the Banda Arc. Initial collision occurred between the Banda volcanic arc and promontories of the continental margin. The lower plate continues to subduct (roll back) in uncollided regions while the upper plate remains locked to the lower plate at the initial points of collision. This ssz spreading produces oceanic crust above the dehydrating lower plate. Water from the lower plate imparts an arc signature on the new lithosphere (Pearce 1982). The development by transpression and transtension of small rhombochasmic basins in the forearc collision zone is a similar mechanism which account for many of the features of Tethyan-type ophiolites.

(b) General case of origin and emplacement of Tethyan-type ssz ophiolites

This is a large issue that obviously cannot be comprehensively covered in a reply to discussion but Dr Clift has focused his question on the apparent lack of associated well-developed arcs with many such ophiolites. The key words here, we believe, are 'apparent lack' and 'well developed'; and later Dr Clift uses the expression 'in almost no instance', weasel words perhaps, how many examples are needed to demonstrate a mechanism can occur? The example of the Banda Arc reveals that even in very young (3 Ma) collision zones any allochthonous volcanic arc terranes may be much broken up by low-angle normal extension faults and steep-angle normal faults, maybe highly dissected and deeply eroded, so much so that all that remains exposed on Timor of the allochthonous Cretaceous–Eocene Sumba–Palelo Arc are several scattered plutons and slices of forearc.

However, some parts of the forearc of the associated volcanic arc in the collision zone have not been overthrust, but on the contrary have been overridden by the converging Australian continental margin. We can see in the region north of the southern part of the Banda Arc between Savu and Tanimbar islands, especially where the exposed Australian margin (Timor island) comes to within 20 km of the exposed Banda volcanic Arc (Atauru island), that part of the forearc having surface dimensions 750 km long by about 150 km wide has 'disappeared'. If the forearc can be overridden, why not the arc? There are only two explanations for the disappearance of this large slice of Banda forearc lithosphere during the last 3 Ma, either it has been overridden by the Australian continental margin as Price & Audley-Charles (1983, 1987) suggested, or it must have slipped laterally along the forearc by major strike–slip faults parallel to the arc. Partly because it has not been found anywhere at the surface Price & Audley-Charles (1983, 1987) proposed it had been underthrust southwards below northern Timor.

The problem with missing arcs associated with orogenic zones is a long-standing argument. Volcanic arc detritus is a common component of most synorogenic deposits in orogenic belts such as the Alps, Cordillera, Oman, and Brooks Range. Sedimentologists find the detritus and trust in heaven for the source. Palaeogeographers see evidence of trenches but very few arc terranes. The evidence suggests that arcs rarely come out of orogens the same way they went in, but their existence cannot be denied. Mitchell (1983) provided a thorough discussion of the problem of missing arcs, suggesting large-scale backthrusting as a mechanism. However, this mechanism alone is inadequate in that these overthrust or buried arc terranes should turn up somewhere in older, more deeply eroded terranes. In some old mountain belts remnants are found, but if, in addition to overthrusting, the arcs are tectonically eroded by continued subduction or especially subduction polarity reversal, then it would be highly unlikely to find a trace of even 'well-developed' arc systems.

The best example of a vanished or 'Houdini' arc system is the northern Luzon volcanic arc in collision with Taiwan. In southern Taiwan, where arc–continent collision is beginning, the volcanic arc is separated from the fold-thrust belt by a narrow remnant forearc basin, comparable with the Wetar Strait of the Banda Arc. The forearc basin and Luzon volcanic arc remnants progressively narrow to the north where the oblique collision becomes progressively older. North of Haulien, where the collision is around 4–6 Ma (ancient) the polarity of subduction has reversed and no surface expression of the arc remains. The bathymetry of this region suggests most of the arc has been devoured by subduction jaws.

These same jaws are in the process of opening along the Wetar thrust, which represents one of the few places in the world where the critical dynamic relations associated with subduction initiation can be observed. The thrust most likely initiated in arc crust and therefore some of the Banda volcanic Arc has already disappeared (Price & Audley-Charles 1987). If this new subduction system develops, as did the Ryukyu trench in Taiwan, perhaps no trace of the Banda volcanic Arc will remain by the time SE Sulawesi and Timor collide. This throws doubt on Dr Clift's theoretical arguments about density contrasts prohibiting the subduction of arc crust.

Dr Clift suggests that ophiolites are generated and 'soled' after the initiation of intraoceanic subduction. Boudier & Coleman (1981) and more recently Boudier & Nicolas (1988) have championed the intra-oceanic subduction model, suggesting the thermal instabilities at a ridge are an ideal site for subduction initiation. The entire mid-Atlantic ridge and East Pacific rise argue against this suggestion. However, the most compelling evidence against the idea, and probably the reason why Coleman has subsequently abandoned it, is the composition of the ophiolite sole. The argument against this mechanism is simple; if an ophiolite was detached at a ridge or some other intra-oceanic setting its composition would be similar to that of its metamorphic sole (the rocks first accreted to the base of the hot slab). The fact is that metamorphic soles differ significantly from the ophiolite that underlie them. Some of the most compelling compositional differences involve the early incorporation of continental-type sediments into the sole with cooling ages that overlap with ages from the ophiolite and synorogenic sedimentation. The mafic igneous rocks at the base of most ophiolites differ petrologically and geochemically from the ophiolitic igneous sequences.

Most of Dr Clift's question deals with concepts addressed at length for years by geoscientists from various perspectives. The final sentence of his question conflicts with the results of the comparative study of ophiolite emplacements (Harris 1989), and those of several others as far back as Hess (1955), who inferred the close ties in time and space between ophiolite development and continental margin shortening. Searle & Stevens (1984), and others who have studied emplacement of ophiolites, such as Moores (1984) and Edelman (1988), all agree with Hess's inferences based on the mountain of radiometric geochemical and stratigraphical data now available from several detailed studies of Tethyan-type ophiolites. The relationships are clear in suggesting these ophiolites are synorogenic and kinematically linked to progressive contraction of old continental margins.

Dr Clift's figure of 100 Ma between ophiolite emplacement and a major collision event is almost two orders of magnitude greater than all the time elapsed so far during the Banda

Arc–Australian margin collision. During those 3 Ma nappes have moved 50 km over the Australian margin, the collision zone has been uplifted at least 5 km (more likely 6 km), and subduction direction in the Timor region appears to have reversed. We accept that during Dr Clift's 100 Ma tectonic activity might have been intermittent. Nevertheless, experience of the 3 Ma old Banda Arc–continent collision zone teaches us that 1 Ma is a long time in collision tectonics and 3 Ma is such a long time that many large-scale events can be completed.

Additional references

Boudier, F. & Coleman, R. G. 1981 Cross-section through the periodotites in the Semail ophiolites, S. E. Oman. *J. geophys. Res.* **86**, 2573–2592.

Boudier, F. & Nicolas, A. 1988 The ophiolites of Oman. *Tectonophysics* **151**. (401 pages.)

Edeleman, S. H. 1988 Ophiolite generation and emplacement by rapid subduction hinge-retreat on a continent-bearing plate. *Geology* **16**, 311–313.

Hess, H. H. 1955 Serpentines, orogeny and epecrogeny. In *Crust of the Earth* (ed. A. Poldervaart). *Geol. Soc. Am. Spec. Paper* **62**, 391–408.

Mitchell, A. H. G. 1983 Where have all the old arcs gone? *Philippine Geologist* **37**, 20–41.

Moores, E. M. 1984 Model for the origin of the Troodos massif, Cyprus and other mideast ophiolites. *Geology* **12**, 500–503.

Pearce, J. A. 1982 Trace element characteristics of lavas from destructive plate boundaries. In *Andesites* (ed. R. S. Thorpe), pp. 339–349. Wiley.

Price, N. J. & Audley-Charles, M. G. 1983 Plate rupture by hydraulic fracture resulting in overthrusting. *Nature, Lond.* **306**, 572–575.

Searle, M. P. & Stevens, R. K. 1984 Obduction processes in ancient, modern and future ophiolites. In *Ophiolites and Ocean Lithosphere* (ed. I. G. Gass, S. J. Lippard & A. W. Shelton) *Geol. Soc. Lond. Spec. Publ.* **13**, 303–320.

Palaeomagnetism, North China and South China collision, and the Tan-Lu fault

By Jin-Lu Lin and M. Fuller

Department of Geological Sciences, University of California, Santa Barbara, California 93106, U.S.A.

Refined Apparent Polar Wander (APW) paths for the North and South China Blocks (NCB and SCB) are presented and the collision between the NCB and SCB discussed. We suggest that the amalgamation of the NCB and SCB was completed in the late Triassic–early Jurassic, during the Indosinian Orogeny. This proposed timing is based on an analysis of palaeomagnetic signatures relating to continental collisions, such as the convergence of palaeolatitude, deflections of declination, hairpin-like loops in and superposition of APW paths. Like the Cenozoic India–Eurasia collision, the Mesozoic NCB–SCB collision reactivated ancient faults in eastern China, converting some of them into transcurrent faults, of which the Tan-Lu fault is the most famous.

Introduction

The NCB and SCB are two of the major blocks in eastern Asia (figure 1). It is now generally agreed that these two blocks were separated from each other in the geological past, and that they collided later to give rise to the Qinling foldbelt. However, the question of when they finally amalgamated has been a matter of debate (Mattuaer *et al.* 1985; Sengor 1985; Hsu *et al.* 1987; Lin *et al.* 1985; Zhao & Coe, 1987). The Tan-Lu fault is a major transcurrent fault in eastern Asia on which a large displacement took place in the early Mesozoic (Xu *et al.* 1987), but the mechanism for its reactivation and displacement has not been clarified.

In this paper we discuss the palaeomagnetic data from the NCB and SCB, examine the palaeomagnetic signatures relating to continental collision, and provide palaeomagnetic constraints on the timing and style of the NCB–SCB collision. Finally, we discuss the relationship between the NCB–SCB collision and the Tan-Lu fault.

Palaeomagnetic data

All presently available palaeomagnetic results for the NCB and SCB are summarized in table 1, and their APW paths are depicted in figure 2. Detailed documentation of the database used here has been published in Chinese journals (Lin 1987, 1989). The new data for the NCB and the new Mesozoic data for the SCB reported by several groups are generally in agreement with each other, and also in agreement with the preliminary APW paths (Lin *et al.* 1985). Here we discuss only the controversial Permian and Tertiary results, and comment on a recently published late Triassic result from the SCB (Zhu *et al.* 1988).

A recent study of some Palaeocene–Eocene sediments yields a pole close to the present north pole (Zhuang *et al.* 1988). This is in conflict with the previous Neogene pole. Moreover, this previous Neogene pole gives rise to a kink in the Neogene APW path. This kink contradicts the generally accepted notion that APW paths are composed of a series of small circles, such as we

133

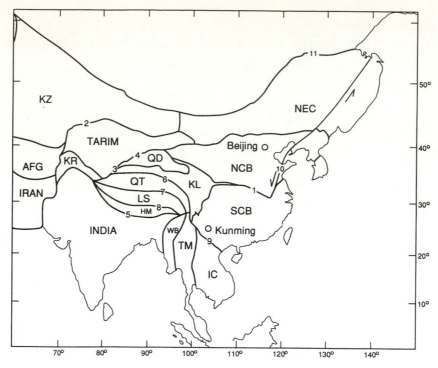

FIGURE 1. Blocks and terranes in China and East Asia (Mercator projection). Abbreviations are as follows. AFG, Afghanistan Terrane; HM, Himalayan Terrane; IC, Indochina Block; IRAN, Central Iran Terrane; KL, Kunlun Terrane; KR, Karakoram Terrane; KZ, Kazakhstan Block; LS, Lhasa Terrane; NCB, North China Block; NEC, Northeast China composite terrane; QD, Qaidam Terrane; QT, Qiangtang Terrane; SCB, South China Block; TM, Thai-Malay Terrane; WB, West Burma Terrane. 1. Qinling-Dabie suture; 2. Tien Shan Mountains; 3. Kunlun Mountains; 4. Altyn Tagh fault; 5. Main boundary thrust (MBT); 6. Jinsha suture; 7. Banggong–Nujiang suture; 8. Indus-Zangpo suture; 9. Red River fault. 10. Tan-Lu fault; 11. Mongol–Okhotsk foldbelt.

see on the APW paths for North America and northern Eurasia, which are the two best defined APW paths at present (Gordon et al. 1984). As the earlier Neogene pole is based only on three lava flows and an early result from redbeds which were subjected solely to alternating field (AF) demagnetization, we prefer to exclude our Neogene pole from the APW path until additional data justify its inclusion.

Zhu et al. (1988) have reported late Triassic results (table 2), which exhibit pronounced streaking along a small circle about the sampling localities, and clearly do not have a fisherian distribution (figure 3). These results differ from each other mainly in declination, but not in inclination. This probably reflects differential rotations at various sampling localities. A statistical analysis of inclination alone based on the method of McFadden & Reid (1982) yields a mean palaeolatitude of $20.1° \pm 2.3°$. When a small circle with a radius of $69.9°$ is drawn about the average sampling centre ($27°$ N, $102°$ E), it intersects with the APW path for the SCB, and the intersection point falls between the middle Triassic and middle Jurassic poles (figure 3). We suggest that it is not the overall mean (mean A in table 2), but the mean B which is calculated on the palaeopoles close to the intersection point, most likely represents the late Triassic palaeopole position of the SCB.

The Permian result from the Emeishan basalts by McElhinny et al. (1981) was initially not included in the preliminary APW path for the SCB because of the possibility of remagnetization.

134

TABLE 1. PALAEOMAGNETIC DATA FROM THE NORTH AND SOUTH CHINA BLOCKS

no.	age	Ma	palaeopole lat. N	lon. E	K	A95	B	Test	palaeolatitude Ob	Ex

A. The South China Block

no.	age	Ma	lat. N	lon. E	K	A95	B	Test	Kunming	Beijing
1	N	10	74.2	36.5	148	7.6	8	R	30.3	41.0
2	E1–E2	53	85.2	174.6	46	(11.9)	6	R	26.4	42.4
3	K2–E1	65	76.3	172.6	23	10.3	10	F	29.0	46.4
4	K1–Km	100	67.6	205.1	35	5.8	19	FR	18.6	36.8
5	K1	125	76.2	225.7	662	4.8	[3]	R	17.1	34.2
6	J3	140	73.0	213.7	29	12.6	6		18.1	35.8
7	J2	165	71.5	201.1	135	5.8	6	R	21.1	39.1
8	Tr3	215	63.9	198.4	530	3.3	5	F	20.0	38.4
9	Tr2	235	54.6	209.7	95	(5.7)	2	RF	11.2	29.7
10	P2–Tr1	250	48.4	219.7	215	4.6	[6]	FR	2.6	21.1
11	P	260	28.5	226.6	112	11.7	3	R	−13.8	4.0
12	C	320	22.2	223.8	83	8.5	5	R	−15.7	1.5
13	D2	380	−8.9	190.4	13	6.9	7	F	−1.4	6.0
14	S2	425	6.8	195.6	122	4.0	12	F	0.5	12.3
15	Є1	560	3.4	195.0	28	8.8	8	RF	−0.4	10.6

B. The North China Block

no.	age	Ma	lat. N	lon. E	K	A95	B	Test	Beijing	
1	N	20	80.6	183.1	91	7.1	6	R	43.1	
2	K2	80	69.0	182.0	193	4.0	8		45.4	
3	K1	115	69.0	200.0	40	12.2	5	R	39.0	
4	J3	140	72.8	214.6	426	6.0	[3]		35.5	
5	J2	165	81.0	238.0	194	8.9	[3]		34.8	
6	Tr2	235	48.9	22.1	418	12.3	[2]	R	26.8	
7	P2–Tr1	250	38.6	9.0	339	5.0	[4]	RF	13.0	
8	P2	255	48.8	0.9	176	9.3	[3]	R	15.6	
9	O2	455	43.2	332.5	53	(10.6)	1		−0.5	
10	O	475	29.1	305.8	18	18.6	5	R	−20.3	
11	Є1	560	15.0	298.6	38	9.9	7	R	−34.9	
12	Z2	570	16.5	301.1	35	9.5	8	R	−33.3	

(Geological time abbreviations: N, Neogene; E, Eogene (Palaeogene); K, Cretaceous; J, Jurassic; Tr, Triassic; P, Permian; C, Carboniferous; D, Devonian; S, Silurian; O, Ordovician; Є, Cambrian; Z, Sinian (*ca.* 850–570 Ma). Early, middle and late epochs of geological periods are denoted with subscripts 1, 2, and 3 respectively. Numerical ages in million years (Ma) are estimated ages obtained by referring the relative stratigraphic levels of sampling formations to the Geological Time Scale (Snelling 1987). K is the precision parameter of the Fisher statistics. A95 is the radius of 95% error circle about the mean palaeopole, or the semi-angle of 95% error about the mean direction when it is in parentheses. B is the number of sites, or number of localities when in brackets. In the test column, R represents the reversal test, F the fold test, and C the contact test. In addition to the observed palaeolatitudes (Ob) of Kunming and Beijing, the expected palaeolatitude (Ex) of Beijing is also calculated relative to the palaeopoles of the SCB. The present geographic coordinates of the two chosen reference points are: Beijing (40° N, 116° E), and Kunming (25° N, 103° E). Sources of data: A. the South China Block: no. 1. Lin *et al.* (1985); 2. Zhuang *et al.* (1988); 3. Kent *et al.* (1986); 4. Lee *et al.* (1987); 5–7. Lin *et al.* (1985); 8. Zhu *et al.* (1988); 9. Chan *et al.* (1984); 10. Lin *et al.* (1985), Opdyke *et al.* (1986), Zhang (1984), Li *et al.* (1989), Steiner *et al.* (1989); 11. Lin (1987 *b*), Lin *et al.* (1985), Zhou *et al.* (1986); 12. Lin *et al.* (1985); 13. Fang *et al.* (1989); 14. Opdyke *et al.* (1987); 15. Lin *et al.* (1985); B. the North China Block: no. 1. Lin *et al.* (1985), Zhao (1987); 2, 3. Lin *et al.* (1985); 4. Lin (1987 *a*); 5. Lin (1987 *a*), Cheng *et al.* (1988), Fang *et al.* (1988); 6. Cheng *et al.* (1988), Fang *et al.* (1988); 7. Lin (1987 *a*), Cheng *et al.* (1988), McElhinny *et al.* (1981); 8. McElhinny *et al.* (1981), Zhao and Coe (1987); 9. Lin *et al.* (1985); 10. Zhao (1987); 11, 12. Lin *et al.* (1985).)

Several new studies have been carried out since then. A recent reinterpretation of these data proposes a predominant normal or mixed polarity of these mid and late Permian basalts (Zhou *et al.* 1986; Zhao & Coe 1987). However, the normal polarity directions from the Emeishan basalts are questionable for the following reasons. (1) The natural remanent magnetization

135

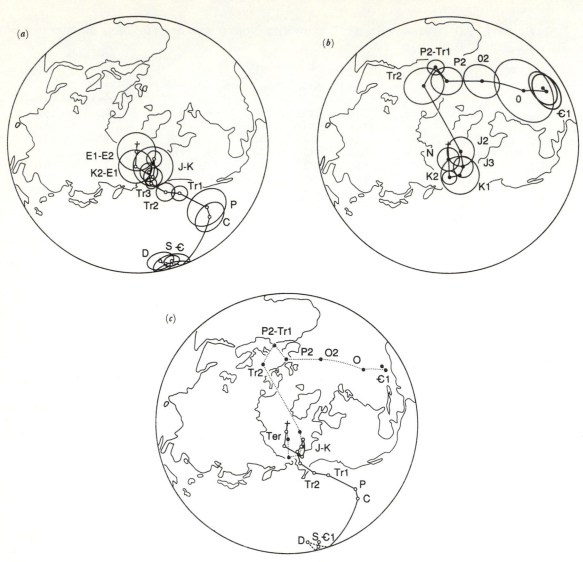

FIGURE 2. The APW paths for (a) South China Block, and (b) for North China Block, and (c) their superposition. Open circles represent palaeopoles from SCB, whereas solid circles those from NCB. Jurassic and Cretaceous poles from SCB are collectively denoted by J-K. Tertiary poles are indicated by Ter. See table 1 for all other abbreviations of geological ages.

(NRM) of the basalts from the Emeishan section consists of a single component of which the *in situ* direction ($D/I = 2.3°/29.2°$) is close to the local dipole field direction (McElhinny *et al.* 1981). (2) The normal and reversed directions are not antipodal but 157.5° away from each other, so that they fail a reversal test (Lin 1989). (3) Although there is a positive fold test, as the folding took place in the latest Cretaceous–Tertiary, so that a Mesozoic remagnetization cannot be precluded. (4) Palaeontological stratigraphy places the Emeishan basalts in the mid-Permian and not latest Permian. The frequently cited isotopic age of 237 Ma of the basalts, if accepted, brings the stratigraphic level of the Emeishan basalts up to the middle Triassic, in evident contradiction with the Permian index fossils in the limestones which are sandwiched by the Emeishan basalts. (5) Steiner *et al.* (1989) have recently obtained some new results from four biostratigraphically well studied uppermost Permian–lowest Triassic sections in eastern

136

FIGURE 3. Late Triassic palaeopoles from western South China. S.L., sampling locality. Small dot represents palaeopole from each site. O.M., the overall mean (triangle) of the 13 site-palaeopoles. Part of the APW path for the SCB is also shown. Notice that the 13 palaeopoles fall on a small circle centred at the sampling locality and with a radius of 69.9°. Also notice that this small circle intersects the APW path between the middle Triassic and middle Jurassic poles. We suggest that it is this intersection point, rather than the fisherian overall mean, which more likely gives the late Triassic pole position.

TABLE 2. UPPER TRIASSIC RESULTS FROM WESTERN SICHUAN AND CENTRAL YUNNAN

no.	locality lat. N	lon. E	N	direction dec.	inc.	K	a95	palaeopole lat. N	lon. E	dp	dm
1*	28.3	103.0	7	28.6	41.3	13	14.6	63.9	198.0	10.8	17.7
2	—	—	6	40.6	43.6	10	18.0	53.9	188.5	14.0	22.5
3	26.6	101.7	6	61.4	36.1	80	6.4	33.7	184.3	4.3	7.4
4	—	—	6	48.2	37.2	11	17.6	45.7	187.8	12.0	20.7
5	—	—	5	59.9	40.2	35	10.6	36.0	181.7	7.7	12.8
6*	26.8	101.5	6	27.4	37.5	11	17.6	64.4	197.6	12.2	20.8
7	—	—	5	35.8	38.2	9	20.9	57.0	193.7	15.2	25.6
8	—	—	5	34.7	37.3	66	7.7	57.8	195.1	5.3	9.0
9	26.4	102.4	6	41.5	35.6	35	9.6	51.4	192.8	6.4	11.1
10	—	—	6	49.5	39.4	30	10.4	45.0	186.3	7.4	12.4
11*	25.1	101.9	6	34.5	22.3	119	5.2	68.4	204.7	3.4	6.0
12*	—	—	5	37.6	29.3	41	9.8	62.8	194.7	6.8	11.5
13*	—	—	5	34.2	32.1	64	8.7	59.6	198.1	5.7	9.9
A.	overall mean		13	—	—	49	6.0	54.0	191.1	—	—
B.	mean (*)		5	—	—	530	3.3	63.9	198.4	—	—
C.	mean		13	—	36.2	106	3.4	palaeolatitude = 20.1° ± 2.3°			

(Lat. N, northern latitude; lon. E, eastern longitude; dec., declination; inc., inclination; N, the number of samples; K, the precision parameter of the Fisher statistics; a95, the half-angle of 95% confidence cone about mean direction; dp and dm, the semi-axes of 95% polar error. No. 1–10 are from western Sichuan, and 11–13 from central Yunnan. No. 1, 2 are from the Xujiahe Formation, 3–5 from the Bingnan Formation, and 6–10 from the Baigewan formation. Results from these three formations pass the general fold test (or tectonic test) at 95%. No. 11–13 are from the Yipinglang area. All samples are from sedimentary rocks. Three means are given in the table, where A is the overall mean, B is a selective mean based on five sites (1, 6, 11–13), and C is a mean based on inclinations only (McFadden & Reid 1982). The original data are from Zhu et al. (1988).)

137

Sichuan, some 200 km away from the Emeishan basalt section. By combining their results with those of Opdyke *et al.* (1986), they give a mean palaeopole at 46.1° N, 223.6° E for the latest Permian–earliest Triassic, which is about 20° away from the palaeopole (49° N, 251° E) for the time interval suggested by Zhao & Coe (1987). Moreover, including the late Permian pole calculated from the normal polarity directions destroys the originally curvilinear feature of the APW path from the Permo–Carboniferous to the late Triassic, and gives rise to a kink in the path. Using the Permian pole in table 1 which is calculated from the reversed polarity direction alone (Lin *et al.* 1985; Zhou *et al.* 1986; Lin 1987 *b*), no kink appears on the APW path. A more detailed discussion of this point has recently been given by Lin (1989).

PALAEOMAGNETIC SIGNATURES OF CONTINENTAL COLLISIONS

Palaeomagnetic signatures of continental collisions include: (1) palaeolatitudinal convergence of colliding blocks, (2) deflection of remanent declinations, (3) hairpin-like loops on the APW paths for colliding blocks, and (4) convergence of APW paths upon collision followed by superposition of the paths after complete amalgamation of colliding blocks. These palaeomagnetic signatures, especially when used collectively, provide important constraints on the timing and tectonic style of continental collisions.

Palaeomagnetism can be most readily applied to the study of the timing and style of continental collisions when the following conditions are satisfied. (1) The collision has taken place within the time for which high-resolution palaeomagnetic results can be obtained. For example, the last collision between North America and Europe took place in the Devonian and led to the Caledonian Orogeny. As the early Palaeozoic palaeomagnetic data from the two continents are still somewhat controversial, it makes a detailed study of palaeomagnetic signature of the Caledonian Orogeny difficult. (2) The two colliding blocks should be large enough in size to provide relatively complete stratigraphic sequences so that sufficiently detailed APW paths can be constructed. The terranes defined in Northwest America do not usually provide complete APW paths, say, since the early Mesozoic (Howell 1985), so that it is hard to unravel collisional history. (3) The post-collisional deformation within the continents and along the collisional boundary should not be so severe that the pre- and syn-collisional palaeomagnetic signatures are totally destroyed by the later deformation. For example, the India–Eurasia collision is difficult to analyse, because there is a large amount of post-collisional crustal shortening in the Tibetan Plateau and along the main central thrust (MCT) and main boundary thrust (MBT). In this case, the shortening along the MBT alone is estimated as 300–500 km or even 800 km (Dewey *et al.* 1988). (4) The necessary palaeomagnetic data must be available from the colliding blocks. (5) The geology and tectonics of colliding blocks should be well documented so that a cross check of geology and palaeomagnetism is possible.

Judging by these five criteria, the collision between the NCB and SCB is suitable for analysis by palaeomagnetic signatures. The SCB contains a complete Phanerozoic sequence, as does the NCB with the exception of the hiatus from the upper Ordovician to the lower Carboniferous. It has been suggested that the collision between them occurred during the Indosinian Orogeny, in the late Triassic and early Jurassic. Palaeomagnetic data for the Permian, Triassic, Jurassic, Cretaceous and Tertiary from the two blocks are available and generally considered reliable, so that the post-Carboniferous APW paths are available for both blocks. The post-collisional deformation within the two blocks and along their suture zone, the Qinling foldbelt, is far less severe than in Tibet and the Himalayas. Finally, the NCB and SCB and the Qinling foldbelt are *terra cognita* geologically. This encourages us to choose the NCB and SCB as a test site to examine the value of the proposed palaeomagnetic signatures in analysis of collision and to gain constraints on the timing and tectonic style of the NCB–SCB collision.

PALAEOLATITUDINAL CHANGE

Figure 4 shows the palaeolatitudinal change of the NCB and SCB, with the cities of Beijing and Kunming chosen as reference sites for the NCB and SCB, respectively. We define the *observed palaeolatitude* for a site, e.g. Beijing, as the latitude which is calculated based on a palaeopole from the block (NCB) on which the site is located, and define the *expected palaeolatitude* as the one calculated from the palaeopole of the other block (SCB). If there had been no relative motion between the NCB and SCB, the expected and observed palaeolatitudes of Beijing should be consistent with each other. On the other hand, if the two palaeolatitude values are different, this indicates that significant relative poleward motion has taken place between the two blocks.

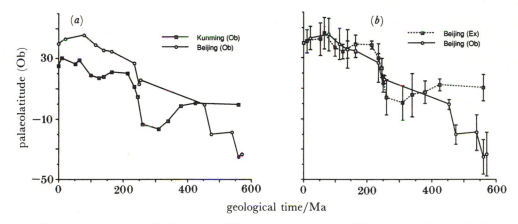

FIGURE 4. (*a*) Observed palaeolatitudinal changes of Beijing in the NCB and of Kunming in the SCB. (*b*) Comparison of the observed and expected palaeolatitudinal values of Beijing. Dashed line represents the expected palaeolatitude of Beijing relative to the palaeopoles of the SCB. Vertical bars are 95% confidence intervals (A95). The two curves are quite different in the early Palaeozoic, but become virtually indistinguishable after the beginning of the Triassic. Plots are based on the data listed in table 1. Palaeomagnetic data are not available for the NCB in the Silurian, Devonian and Carboniferous.

We see in figure 4 that the observed and expected palaeolatitudinal curves of Beijing are superposed since Triassic, but differ significantly in the early Palaeozoic and also in the Permian. This indicates that there has been considerable relative motion between the two blocks since Cambro–Ordovician time. The palaeolatitudes approached each other in the late Permian, became similar in the Triassic. No significant relative poleward motion has taken place since then. These observations suggest that the collision between the NCB and SCB occurred in the Triassic, or the so-called Indosinian Orogeny. Unfortunately, the above interpretation is weakened by the lack of palaeomagnetic results of the Carboniferous, Devonian and Silurian times from the NCB. We therefore turn to other palaeomagnetic signatures.

Rapid deflection of remanent declination

Continental collisions are likely to start at promontories because they form irregularities in the continental outlines (see, for example, Wilson 1966). The two plates may then rotate towards each other as the collision proceeds, or the promontories may rotate to become parallel to the main collision front. Thus collisions are likely to lead to rapid rotations, which give rise to rapid changes in declination without accompanying changes in inclination. In an APW path this gives rise to an arc of a small circle about the site. As we mentioned above, from the Triassic the observed and expected palaeolatitudes of Beijing become consistent with each other, so that the relative change in inclinations of the two blocks essentially stopped. However, a big change in the declination of the NCB started shortly after in the late Triassic. The declination of the early and middle Triassic is about N45° W, but becomes N15° E in the middle Jurassic, indicating 60° counterclockwise (CCW) rotation of the remanent magnetization vector. We interpret this rotation to be related to the collision between the NCB and SCB. We speculate that before the collision the NCB, together with Mongolia–Northeast China, formed a promontory to the south of the Siberian Platform, perhaps similar to the present Malayian peninsula relative to the Asian continent. With the collision of the SCB, the tectonic assemblage of the NCB–Mongolia–Northeast China rotated CCW by 60° into approximately its present configuration with respect to Siberia. This finally closed the Mongol–Okhotsk remnant ocean between Mongolia–Northeast China and the Siberian Platform. There is no comparable rotation recorded for the SCB during the Triassic–Jurassic time. The NCB and SCB must therefore have been still moving somewhat independently at that time and did not form a coherent tectonic entity in middle Triassic time, although their initial collision had probably started by then.

Smooth tracks and hairpin-like loops in APW paths

APW paths often consist of curvilinear arcs, or tracks, and hairpin-like loops. Tracks indicate uniform, relatively rapid plate motion whereas loops indicate stagnation of polar wander, suggesting confrontation with obstacles in plate motion, most likely, continental collisions (Irving & Park 1972; Gordon *et al.* 1984). The APW path for the SCB has a substantial linear segment from the late Carboniferous to the late Triassic, representing rapid and uniform APW during that time. However, this APW was halted during the late Triassic and is much less rapid in the Jurassic and Cretaceous. There may be a loop in the APW path, although the detailed form of the curve is beyond the resolution of the available data. We interpret this Jurassic–Cretaceous loop, or stagnation point, on the APW path for the SCB as evidence for a continental collision. We suggest that the collision was with the NCB.

Superposition of APW paths

The two APW paths for the NCB and SCB are plotted on the same map in figure 2c. They are widely separated in the Palaeozoic, start to approach each other in the early Mesozoic, meet in the middle Jurassic and are indistinguishable subsequently. This indicates that the amalgamation of the two blocks was not accomplished until as late as the early Jurassic, which again indicates an Indosinian collision.

NCB–SCB COLLISION AND DISPLACEMENT ON THE TAN-LU FAULT

Continental collisions may reactivate ancient faults in continents, sometimes converting them into transcurrent faults on which a large amount of strike–slip motion takes place. The Cenozoic tectonics in Asia following the India–Eurasia collision affords the best known example of this behaviour (Molnar & Tapponnier 1975). In the eastern continental margin of Asia the Tan-Lu is the largest and most important fault on which a major displacement started in the late Triassic, i.e. during the Indosinian Orogeny (Xu *et al.* 1987). We suggest that the reactivation of, and displacement on, the Tan-Lu fault is directly related to the collision between NCB and SCB. The present trends of the Tan-Lu fault and the Qinling suture zone between NCB and SCB are nearly orthogonal (figure 1). However, on a middle Triassic map reconstructed based on palaeomagnetic data, the strike of the Tan-Lu fault becomes nearly parallel to the relative motion vector of SCB upon its collision with NCB (figure 5). This palaeogeometry suggests that due to the continental collision and continued northward drive of SCB after the collision, ancient faults in eastern NCB, such as the Tan-Lu fault, were reactivated or converted into left-lateral strike–slip faults.

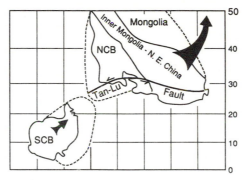

FIGURE 5. A middle Triassic reconstruction map of SCB and NCB, showing their palaeo-orientations as well as their palaeolatitudes. Notice that the northern margin of SCB was parallel to the incipient Tan-Lu fault. This suggests that there may be a relation between the SCB–NCB collision in the one hand, and the reactivation of and displacement on the Tan-Lu fault in the other hand. We suggest that the push from SCB may be strong enough to force the assemblage of NCB–Mongolia–Northeast China to have rotated counterclockwise and thus to have brought the Mongol–Okhotsk palaeo-ocean into final closure.

CONCLUSION

In conclusion, we note that a variety of collision-related palaeomagnetic signatures suggest an Indosinian collision between NCB and SCB. The palaeolatitudes of NCB and SCB converge in the Triassic. There is a rapid change in declination of up to 60° of NCB during the late Triassic and early Jurassic. In the two APW paths there is a Jurassic–Cretaceous hairpin-like loop or stagnation point. Finally, post-Middle Jurassic APW paths for the two blocks are indistinguishable. All of these palaeomagnetic signatures suggest that the collision between the NCB and SCB started in the early or middle Triassic time, and was completed in the late Triassic–early Jurassic time. This is consistent with various lines of geological evidence (Hsu *et al.* 1987), such as the characteristic stratigraphic sequences in the Qinling foldbelt, the Sm–Nd age determination of the collision-related C-type eclogite (Li *et al.* 1989), the radiometric age dating of granitic batholiths in the Qinling foldbelt and so on. The collision model which we propose predicts that the collision of the NCB–Mongolia-Northeast China assemblage with Siberia should begin in the west and progress eastward, while that of the SCB with NCB should start towards the east and progress westward.

REFERENCES

Chan, L. S., Wang, C. Y. & Wu, X. Y. 1984 *Geophys. Res. Lett.* **11**, 1157–1160.

Cheng, G., Bai, Y. & Sun, Y. 1988 *Seismol. Geol.* **10**(2), 81–87.

Dewey, J. F., Shackleton, R. M., Chang, Ch. & Sun, Y. 1988 *Phil. Trans. R. S. Lond.* A **327**, 379–413.

Fang, D., Guo, Y., Wang, Zh., Tan, X., Fan, Sh., Yuan, Y., Tang, X. & Wang, B. 1988 *Kexue Tongbao* **33**(2), 133–135.

Fang, W., Van der Voo, R. & Liang, Q. 1989 *Tectonics* **8**, 939–952.

Gordon, R. G., Cox, A. & O'Haire, S. 1984 *Tectonics* **3**, 499–537.

Howell, D. G. (ed.) 1985 *Tectonostratigraphic terranes of the Circum-Pacific region.* Circum-Pacific Council for Energy and Mineral Resources, Houston, Texas, U.S.A. (581 pages.)

Hsu, K. J., Wang, Q., Li, J., Zhou, D. & Sun, Sh. 1987 *Eclogae Geol. Helv.* **80**, 735–752.

Irving, E. & Park, J. K. 1972 *Can. J. Earth Sci.* **9**, 1318–1324.

Kent, D. V., Xu, G., Huang, K., Zhang, W. Y. & Opdyke, N. D. 1986 *Earth planet. Sci. Lett.* **79**, 179–184.

Lee, G., Besse, J., Courtillot, V. & Montigny, R. 1987 *J. geophys. Res.* B **92**, 3580–3596.

Li, H., Wang, J., Heller, F. & Lowrie, W. 1989 *Chin. Sci. Bull.* **34**, 431–434.

Li, Sh., Ge, N., Liu, D., Zhang, Z., Ye, X., Zheng, Sh. & Peng, Ch. 1989 *Kexue Tongbao* **33**, 522–525.

Lin, J. 1987a *Scientia Geol. Sin.* **2**, 183–187.

Lin, J. 1987b *Scientia Geol. Sin.* **4**, 306–315.

Lin, J. 1988 *Seismol. Geol.* **10**(3), 1–11.

Lin, J. 1989 *Geol. Rev.* **35**, 349–354.

Lin, J., Fuller, M. & Zhang, W. 1985 *Nature, Lond.* **313**, 444–449.

Mattauer, M., Matte, Ph., Malavieille, J., Tapponnier, P., Maluski, H., Xu, Zh., Lu, Y. & Tang, Y. 1985 *Nature, Lond.* **317**, 496–500.

McElhinny, M. W. 1985 *J. Geodynamics* **2**, 115–117.

McElhinny, M. W., Embleton, B. J. J., Ma, X. H. & Zhang, Z. K. 1981 *Nature, Lond.* **293**, 212–216.

McFadden, P. L. & Reid, A. B. 1982 *Geophys. Jl R. astr. Soc.* **69**, 307–319.

Molnar, P. & Tapponnier, P. 1975 *Science, Wash.* **189**, 419–426.

Opdyke, N. D., Huang, K., Xu, G., Zhang, W. & Kent, D. V. 1986 *J. geophys. Res.* B **91**, 9553–9568.

Opdyke, N. D., Huang, K., Xu, G., Zhang, W. Y. & Kent, D. V. 1987 *Tectonophysics* **139**, 123–132.

Sengor, A. M. C. 1985 *Nature, Lond.* **318**, 16–17.

Snelling, N. J. 1987 *Modern Geol.* **11**, 365–374.

Steiner, M., Ogg, J., Zhang, Z. & Sun, S. 1989 *J. geophys. Res.* **94**, 7343–7363.

Wilson, J. T. 1966 *Nature, Lond.* **211**, 676–681.

Xu, J., Zhu, G., Tong, W., Cui, K. & Liu, Q. 1987 *Tectonophysics* **134**, 273–310.

Zhang, Zh. 1984 *Bull. Chi. Acad. Geol. Sci.* **9**, 45–54.

Zhao, X. 1987 Ph.D. thesis, University of California at Santa Cruz, U.S.A.

Zhao, X. & Coe, R. C. 1987 *Nature, Lond.* **327**, 141–144.

Zhou, Y., Lu, L. & Zhang, B. 1986 *Geol. Rev.* **32**, 465–469.

Zhu, Zh., Hao, T. & Zhao, H. 1988 *Acta Geophys. Sin.* **31**, 420–431.

Zhuang, Zh., Tian, D., Ma, X., Ren, X., Jiang, X. & Xu, Sh. 1988 *Geophys. Geochem. Explor.* **12**, 224–227.

Terrane provenance and amalgamation: examples from the Caledonides

By B. J. Bluck

Department of Geology and Applied Geology, University of Glasgow, Glasgow G12 8QQ, U.K.

The Scottish Caledonides have grown by the accretion of terranes generated somewhere along the Laurentian margin. By the time these terranes had been emplaced along the Scottish sector, they were structurally truncated then reassembled to form an incomplete collage of indirectly related tectonic elements of a destructive margin. The basement to some of these tectonic elements and the basement blocks belonging to the previously accreted Precambrian are of uncertain provenance and a source in the Pan-African craton is possible.

As terranes migrate along the orogen they generate in the fault zones and on their periphery a reservoir of mature sediment. This mature sediment is produced because of the recycling produced during the generation and destruction of sedimentary basins developing during terrane translation. At each period of recycling the mature sediments are mixed with less mature sediments yielded from local uplifts generated by the new basin formation.

If a large part of the orogen suffers orthogonal closure, giant river systems may form and disperse sediment across terranes. This is likely to have happened during the Devonian–Carboniferous of parts of N. Europe.

1. Introduction

The identification, provenance and amalgamation history of terranes in the Palaeozoic may require the use of techniques which are not normally applied or are inappropriate for use in the younger parts of the stratigraphic record where there are clearly defined faunal provinces and where there may be a comparatively straight-forward palaeomagnetic record. The Palaeozoic has no off-shore record of an ocean spreading history which may otherwise aid in determining the provenance of terrane blocks. Reliance has to be made on other criteria and some of these, as used in the Caledonides, are presented in this paper. In Caledonian terrane analysis there are three aspects which are important: (i) terrane identification (ii) terrane provenance, often in the absence of a meaningful faunal record, and (iii) terrane amalgamation history.

2. Criteria for terrane identification

At least seven terranes make up the Scottish Caledonides (figure 1) and the following criteria have been used to identify them and their boundary zones.

(a) *Incompatible histories.* Where the uplift of a basement block has not resulted in the supply of sediment to a basin now lying adjacent to it when both uplift of the basement and subsidence of the basin are known to have taken place at the same time, then large-scale movement between basement and basin can be inferred.

An example of this can be seen in the juxtaposition of the Highland Border Complex with the Dalradian block (figure 1a). The Highland Border Complex comprises a fragmented

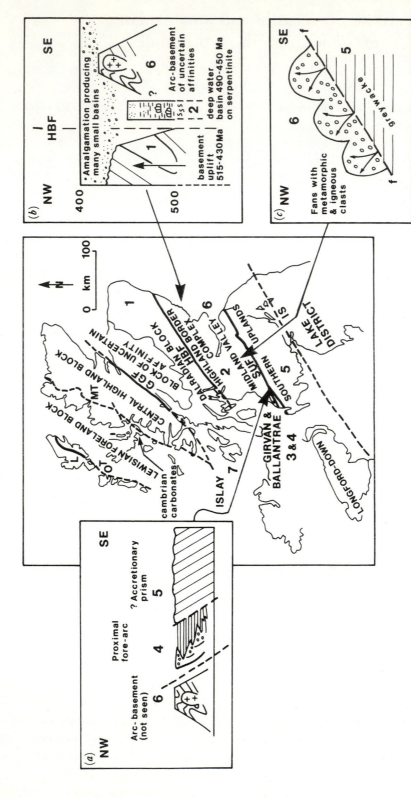

FIGURE 1. Terrane map of the Scottish Caledonides showing some of the criteria by which they have been recognized. 1, Dalradian Block (metamorphic basement); 2, Highland Border Complex (Ophiolitic and other rocks); 3, Ballantrae Complex (Ophiolite); 4, Girvan (forearc); 5, Southern Uplands (accretionary prism?); 6, Midland Valley of Scotland arc-basement; 7, Islay–Colonsay (metamorphic basement). L, Lewis; MT, Moine thrust; GGF, Great Glen Fault; HBF, Highland Boundary Fault; SUF, Southern Uplands Fault; IS, Iapetus Suture. (a) Section showing the incompatible history of events across the Highland Boundary Fault. (b) Section showing loss of tectonic elements across the Southern Uplands Fault. (c) The mismatch of clast type and local potential source across the Southern Upland Fault.

ophiolite, black shales, limestones and lithic-arenites with a provenance in volcanic and ophiolitic sources. It extends as a thin sliver all along the Highland Boundary Fault in mainland Scotland; is thought to be present in the Tyrone inlier in central Northern Ireland and may have as its lateral continuation the Clew Bay sequence in W. Ireland (Harper *et al.* 1989). Deposition of these sediments ranges fairly continuously from Lower to Upper Ordovician but a sliver of Middle Cambrian rocks occur in the region of Callander (Curry *et al.* 1982). The Dalradian block now adjacent to the north underwent uplift and erosion during the late Cambrian–late Ordovician as recorded in a variety of age determinations (Dempster 1985). The Highland Border Complex is neither in a facies nor does it have the type of sediment expected if it were lying adjacent to such a block during the period of its uplift.

(b) *Missing sediment piles.* Large-scaled metamorphic terranes should have associated sediment piles which were generated during their uplift. On the other hand some metamorphic complexes, particularly core-complexes, would have a record of uplift dominated by structurally bounded sheets of cover. Studies of recent mountain belts have shown that these two forms of unroofing are often seen together, and the sediment yielded from such uplifted areas is sometimes a complex mixture derived from various sources and differing levels within the orogen. Sediment may not be dispersed directly to a permanent basin but may be stored in many small, temporary basins which exists within the orogenic zone and so may be polycyclic by the time it reaches a preservable basin. Sediment is seldom a direct and simple record of the source uplift.

A corollary of (a) (above) is that the sediment generated by the late Cambrian–late Ordovician uplift is missing. The Dalradian sequence is seen as a nappe pile diverging from a central root zone of vertical rocks which occupies the ground parallel with and just south of the Great Glen Fault. The nappes, and possibly an early thermal-uplift event, were generated in late-Precambrian times (Rogers *et al.* 1989). To the SE and NW there should have been foreland basins produced during this nappe emplacement episode, but there is no known record of these basin. Neither has the Cambro-Ordovician uplift related to the second phase of Dalradian development left any known sedimentary record to the SE or the NW. No sediment from this uplift is recorded in a coeval limestone sequence of Cambro-Ordovician age to the NW, although these rocks would have been distanced from the areas of Dalradian uplift by the amount of displacement on the Moine Thrust.

(c) *Missing tectonic units.* Terrane accretions on destructive margins are often typified by assemblages of arc-complexes and marginal basins created somewhere along the margin then to be reassembled elsewhere. In that sense they are not exotic to the continent to which they accrete. Each of the elements of a destructive margin are often divided from others by faults or zones of structurally weak rock. This characteristic may allow the preferential fragmentation of the complete destructive margin assemblage along these zones of weakness so that parts of it become detached and move as separate blocks. The reassembling may or may not take place in the order of its generation or some of the tectonic elements may be missing.

Since the Cambrian, most of the tectonic elements which have accreted along the Southern margin of Laurentia have faunas of North American aspect, so they were probably generated somewhere along that margin and then transported elsewhere. Examination of the existing record shows that there are many gaps between tectonic elements, and some are now only recorded as slivers along fault zones. The boundary between the Southern Uplands and the Midland Valley provides an example of ground missing between tectonic elements (figure 1 b).

If it is accepted that the Southern Uplands is an accretionary prism and the Girvan region, to the immediate N of the Southern Uplands Fault, a proximal fore-arc basin, then there is gap between these two tectonic elements normally filled by a fore-arc region which is usually more that 100 km in width (Bluck 1983).

(d) *Incomplete tectonic blocks.* These may be identified where there is a need for the presence of additional crust to explain of the genesis of the existing crust. In this case the folded and metamorphic, Dalradian block is now juxtaposed against the relatively unmetamorphosed Highland Border Complex. Dalradian greenschist facies rocks with vertical structure lie in this contact, but the cover needed to produce the degree of metamorphism in these greenschist rocks is in the order of 10–15 km. This cover is now missing and must have been stipped off before or possibly during the emplacement of the Highland Border Complex.

(e) *Mismatches between compositions of the sediments and the composition of the blocks in the direction of dispersal.* In the final stages of terrane amalgamation there are a number of problems associated with this kind of relationship (as will be discussed later) but in some instances the mismatches are fairly clear. Along the Southern Uplands Fault sediments dispersed from the SE do not contain clasts of the blocks which presently lies in that direction (figure 1c). The sediments, deposited in alluvial fans contain abundant igneous and metamorphic clasts up to boulder size, but in the direction from which they were transported lies the Southern Uplands, a sequence of greywackes and shales. In this instance a source block has been laterally displaced or overthrust by the greywacke sequence (Bluck 1985).

(f) *Identification of cryptic terranes.* A terrane may be detached from the sediment that it sheds. In the case of the Dalradian block, there is an Ordovician sediment source with no certain, identifiable sediment pile related to it. The converse also applies, and this is probably true of much of the sediment which has accumulated in basins around the Midland Valley of Scotland.

The provenance of sediment in the Southern Uplands and the nature of the Southern Uplands Block are both a matter of controversy. The block itself is bounded by the Southern Uplands fault to the north and the Solway-line to the south, and is cut by a prominent fracture, the Orlock Bridge Fault. This fault is thought by Anderson & Oliver (1986) to be a terrane boundary bringing together two blocks of similar lithology but with different geological histories. The tectonic nature of the Southern Uplands, whether it is an accretionary prism, fore-arc basin, back-arc basin or a compound of basins with different affinities have all been discussed at length (McKerrow 1987). In addition, the dispersal within the Southern Uplands is also debated; there are undoubtably flows directed towards the SE, and NW, but the bulk of the sediment is dispersed axially along the trough.

The sediments are mainly lithic-wackes and arenites, together with lutites, mudstones, and some conglomerates. The wackes contain abundant igneous clasts, metamorphic minerals and fragments, and often an abundance of compacted and sometimes sheared shales and mudstones. Although there is a wide range in the dispersal directions for the wackes, the conglomerates, some of which are boulder bearing, are usually dispersed from the NW. Among the clasts contained in these conglomerates are granitoids, volcanic, metamorphic and some sedimentary detritus. The granitoids clasts have cooling ages very near to the estimates of the time of their sedimentation; metamorphic clasts and muscovite micas in the wackes and arenites both have cooling ages which are also near to the depositional age of the sediments containing them (Kelley & Bluck 1989). The isotopic data taken together with the general composition of the sediment suggesting a provenance in an arc terrane with metamorphic basement, probably to the NW.

At present there is no such arc terrane with plutons and associated metamorphic basement to be seen in the ground to the immediate NW; it may be buried under a cover of Upper Palaeozoic sediment now comprising much of the Midland Valley and its extension SW into Ireland. However, there may be grounds for interpreting the metamorphic basement, the acidic intrusive, and extrusive rocks of the Tyrone Complex (Hartley 1933) as *in situ* Midland Valley basement.

3. PROVENANCE OF TERRANES

The signature given by faunas to crustal blocks is probably among the most diagnostic in terms of tracing the provenance of terranes. Faunal provincialism is the great gift to the Phanerozoic terrane tectonist. For the deciphering of terrane provenance before the emergence of faunas and provincialism other, and altogether less satisfactory techniques have to be used.

Precambrian cratonic blocks in the N. Atlantic region at least grew in two main cycles, at roughly 3.0 Ga and 2.0 Ga (Patchett & Arndt 1986); from that last growth event up to the base of the Cambrian, crustal units underwent further deformational and growth periods which gave them a degree of characterization which allowed them in a very general sense to be recognized as distinctive. The release of a basement block into the zone of terrane migration, or the irregular fragmentation of a continent during the initiation of a spreading cycle allows many blocks with a variety of signatures to be available for incorporation into the terrane collage. These blocks may subsequently be coated with a sedimentary record diagnostic of a particular faunal province or may become intruded, covered with extrusives or remetamorphosed in a totally new tectonic régime.

Much of the North American Archaean and early Proterozoic craton was sealed off from the zone of Caledonian terrane migration by the pervasive Grenville (1.2–0.9 Ga) event (figure 2). Blocks entering the Caledonides without Grenville overprinting either did not originate from this part of the craton or escaped via tectonic lines which transected the craton. On the other hand, the Pan-African block was suitable for the release of tectonic blocks without pervasive Grenville overprinting since the Grenville event is poorly represented there and appears not to have pervasively modified the crust at its Atlantic margin.

There is another great advantage that results from periods of rapid crustal growth during the Precambrian. Where the crust has grown rapidly, there is usually a substantial or dominant contribution directly from the mantle, which can be recognized from model Nd age studies (see Patchett & Arndt 1986). The thermal events which may accompany such periods of rapid crustal growth or subsequent intra-crustal episodes are recorded in the radiometric ages of the various mineral phases, each of which blocks at a characteristic temperature. By combining information from both crustal growth processes and from thermal resetting it is possible to constrain the provenance of a suspect crustal block of Precambrian age.

In the Caledonides, clasts of metamorphic rock which occur in the Highland Border Complex, have Rb–Sr cooling ages of *ca.* 1800 Ma, indicating involvement in a 1.8–2.0 Ga thermal event which is usually associated with the crustal growth event of that age. However, model Nd ages of 2.5–2.9 Ga, indicate a provenance in that part of the orogen where sediment of the protolith was derived partly from the Archaean. The block which supplied the sediment had clearly missed any Grenville overprint, at about 1.0 Ga and had not been involved in Pan-

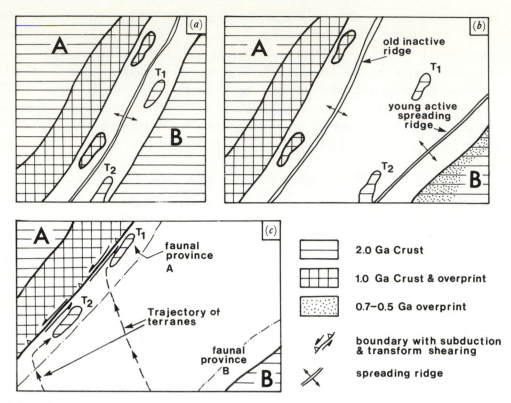

FIGURE 2. Possible paths taken by Precambrian basement terranes during the early phases of ocean opening. Two continents, A (N. America) and B (Pan-Africa) share a history at 2.0 Ga but develop differently from that time on. The terranes generated during initial separation have different histories depending upon where and when they leave the mother continent, and these differences can be read from isotopic data. Whatever their provenance their later history will be dominated by the events on the margin to which they accrete and to recognize the true nature of the terrane one should read through these later events. T_1 and T_2 are terranes and the associated sediment they produce. They came from B but reached A after the establishment of faunal provinces. The boulders and the block from which they came in the Highland Border Complex (Dempster & Bluck 1989) and the Islay–Colonsay terrane may have had this route.

African or Caledonian thermal events. However, by Arenig time (the estimated age of the deposit in which the boulders occur), the block, or sediment yielded from it was in a Laurentian faunal province (Dempster & Bluck 1989).

The possible routes for blocks with this signature are modelled in figure 2, where two continental masses, A (N. America) and B (Africa) are divided by an opening ocean (Iapetus). Both A and B have had a period of accelerated crustal growth at *ca.* 2.0 Ga, but A has, in addition undergone a thermal and crustal accretion event at *ca.* 1.0 Ga (Grenville). The ocean opened parallel and to the right (east) of this later event. During the initial splitting, there are many small continental blocks which are detached from their parent continental masses: those from A carry the record of two widespread crustal events, but those from B (T_1 and T_2) the signature of only one. After the separation of the crustal blocks from B a third thermal event occurred there (Pan-African 1 and 2), but as the continent A and the blocks of B are now some distance away from it they did not take part in this stage of the history of B (figure 2b).

A second spreading ridge developed to the right of the first, and oblique subduction is initiated at the margin of continent A. By this time two important things have happened (i) faunas have developed on the Earth's surface, and (ii) continents A and B are sufficiently far away from each other to permit the development of faunal provinces which become characteristic of the seas bordering each continental mass. Terranes T_1 and T_2 are now being transported on oceanic crust towards continent A where they acquire the signature of the faunal province there as well as the tectono-thermal history that this margin is undergoing and will undergo. In this way Precambrian crustal blocks, in themselves exotic to the continent to which they accrete, may after accretion bear the characteristics which may lead one to assume that they have always belonged to the continental mass in which we now find them.

Although this explanation is offered for some of the boulders in the Highland Border Complex, it may well apply to the Precambrian metamorphic terranes of the Caledonides. It may therefore be argued that the basement blocks which are pre-Caledonian in age, and which may already have accreted to the craton by Cambrian times (i.e. before faunal provinces have been established) could belong to cratons which are now far removed from them. The Islay–Colonsay terrane (figure 1; Bentley *et al.* 1988; Marcantiono *et al.* 1988), for example, appears also to have escaped a Grenville overprint, and as with the terrane which supplied the boulders to the Highland Border, it may have had a similar and complex route to its final docking in Scotland. Rogers *et al.* (1989) have pointed out that because of a change in the time scale of sedimentary and structural events in the Dalradian block its affinities with the Laurentian craton are no longer clear. Indeed the early deformation event in this block being now pre-590 Ma, fall within the time span of the early Pan-African tectono-thermal events, so it is difficult to rule out an origin for this block in Gondwana. In this case a separation would have taken place after the Pan-African events, and the later part of Dalradian history may have taken been acquired within the Laurentian domain.

It follows too that the tectonic and thermal evolution of a terrane need not have taken place in the region where it is at present located. In this context, for example, the Dalradian block may have been undergoing structural and metamorphic developments while moving into its present position.

4. TERRANE AMALGAMATION

The amalgamation history of terranes is partly written in the sedimentary record of terrane borderlands and the plexus of faults surrounding terrane blocks. In both these areas basins are created and destroyed in a very short space of time as the locus of fault activity changes from one sector of the fault zone or branch of a fault to another. In the Betic Cordilleras, for example there have been three periods of basin formation within the past 8–10 Ma, with unconformities separating each period of basin development (Boccaletti *et al.* 1987). Along the Cadiz–Alicante line, a zone of strike–slip activity, basins less than 5 Ma old have been folded and eroded (Boccaletti *et al.* 1987). This rapid basin development and erosion implies a very considerable recycling of sediment. The labile clasts are broken down but the durable lithologies such as quartzite and vein quartz resist degradation to become progressively more rounded. Rather than disappearing from the sedimentary system altogether, these quartzites and vein-quartz clasts tend to form a reservoir of mature sediment even down to sand size which is retained within the fault zones and on the terrane margins which border them. When new basins are opened up, these clasts are then available for redeposition, but now along with the immature sediment which may have been generated by a new uplift accompanying the phase of basin development. In this way highly rounded, mature sediment is mixed with angular labile sediment and both are seen to contribute to sedimentary basins either as complex mixtures or segregated assemblages delivered from one side or the other (figure 3).

The nature of the fill to these sedimentary basins is also fairly diagnostic. They are often filled laterally with coarse sediment derived from alluvial fans bordering the basins. As the basin floor migrates relative to the source overlapping fans are created, and the sense of relative displacement between basin and source is given by the direction of fan overlap (Crowell 1982).

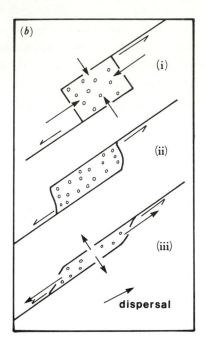

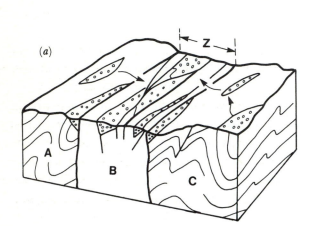

FIGURE 3. (a) The development of basins on the margins of terranes (A and C) and their development in the fault zone (Z); arrows show dispersal, B refers to a suitable lithology, e.g. serpentinite within which the fault zone has developed. (b) Simplified history of a basin formed within the fault zone: (i) development of a pull-apart basin; (ii) stretching and deformation; (iii) inversion of basin to yield detritus to new basin.

Basins with mixtures of mature and immature sediments characterize the stratigraphical sequences which occur along two main terrane boundaries in the Caledonides: the Southern Uplands and Midland Valley Faults. Mature coarse sediments in the form of conglomerates with well-rounded quartzite clasts, appear as early as mid-Ordovician in the Southern Uplands (Floyd 1982) and are also present on the north side of the Midland Valley as well as in the Mayo Trough, but they are particularly characteristic of basins developed during the Old Red Sandstone. Here along the Highland Boundary Fault, basins with overlapping fans are characterized by quartzite clasts, which are often broken and re-rounded, or rimmed by red discolouration both characteristics indicating their involvement in pre-existing cycles of deposition and erosion. In addition some of the sandstones are sublithic arenites which are very rich in rounded and subrounded quartz. Conglomerates rich in quartzite clasts and sandstones rich in quartz grains are sometimes interstratified with lithic arenites which have almost no quartz, and conglomerates which are dominated by subrounded fragments of volcanic or other highly labile rock. The sediments with a mature clast assemblage are often seen to have a different dispersal direction from those dominated by immature sediment. Thick, coarse conglomerates with highly mature clast assemblages imply a source in recycled sediments.

TERRANE PROVENANCE AND AMALGAMATION

When these sediments are interstratified with units with highly immature clast assemblages it suggests that the tectonic and other conditions under which the sediment accumulated were unstable enough to preserve abundant labile debris. The presence of supermature clast assemblages then becomes an anomaly which can easily be explained in terms of a reservoir of mature grains being available for sedimentation in rapidly subsiding basins. In this Silurian–Devonian régime these reservoirs of highly immature sediment along the Southern Uplands and Highland Boundary Fault zones were probably produced by prolonged periods of sediment recycling during the opening and closing of sedimentary basins.

Terrane amalgamation is also recorded other events. Where the orogen is partly or totally the result of orthogonal collision, then high mountains with high sediment yield and high orographic rainfall may produce giant river systems which disperse sediment over other terrane blocks. A classic example is the Himalayan collision with its huge sediment yield and the dispersal of its sediment over a wide area by rivers like the Indus, Ganges, Yellow and Lena. The signature of the collision in this instance is evidently diverse, and dispersed over terranes belonging to more than one orogenic cycle.

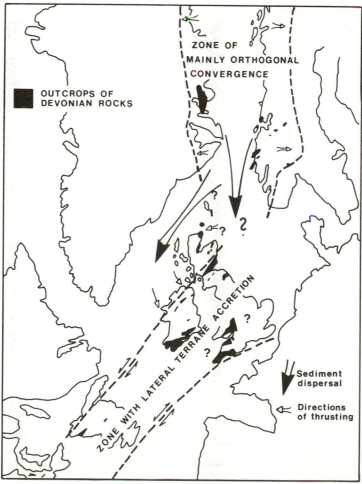

FIGURE 4. Showing the locations of strike–slip and orthogonal collisions during Devonian times. The orthogonal convergence has created high mountains and the sediment dispersed from these mountains dominate the record elsewhere. Small fault basins in the Midland Valley of Scotland are filled by streams which are far larger than one would anticipate from the scale of the basin.

151

In the Caledonian orogen, the strike–slip régime which controlled the development of the British sector during the Old Red Sandstone is replaced to the NW by an orogenic episode characterized by orthogonal collision in the Scandian (Gee 1978) (figure 4). The length of this orogen exceeds 1500 km, and involves Greenland as well as Scandinavia. It is characterized by rapid uplift (Cuthbert *et al.* 1983) and would certainly have been the site of high sediment yield and orographic rainfall. The evidence, in the form of sediment bars greater than 15 m thick, for large river systems in the Old Red Sandstone of the Midland Valley, and the Carboniferous of Northern England (McCabe 1977), together with the widespread occurrence of mica and meta-clast bearing sandstones of Devonian–Carboniferous age suggests a source in this orogen. It also follows that small fault basins may develop in regions such as the Midland Valley of Scotland to be filled by river systems which are far larger than one would anticipate from the size of the basin. In this way the sedimentary fill to these small basins may be dominated by fine sediment from a large distal source rather than the usual coarse sediment from a local source.

5. CONCLUSIONS

Terranes in Caledonian Scotland are truncated fragments of a destructive margin which existed along the edge of Laurentia. Scotland has lost ground during the Caledonian cycle; it is relatively narrow for a destructive margin which has formed in situ. Older crustal blocks, but with Caledonian overprinting in terms of thermal (metamorphic) or igneous events, may not have had their provenance in Laurentia; the Pan-African landmass also provides a more suitable source on the basis of the meagre data available. These may be the truly exotic terranes of the Caledonian. During the latter stages of terrane accretion reservoirs of mature sediment were available for deposition into sedimentary basins, and so mixtures of highly mature and immature sediments were deposited.

Orthogonal collision in Scandinavia may have generated sediments which swept over much of Northern Europe much as sediment from present-day orogens disperse sediment widely from them to accumulate at some distance from their site of genesis.

I thank Dr Nigel Harris for his constructive comments.

REFERENCES

Anderson, T. B. & Oliver, G. J. H. 1986 The Orlock Bridge Fault: a major Late Caledonian sinistral fault in the Southern Uplands terrane. *Trans. R. Soc. Edinb. Earth Sci.* **77**, 203–222.

Bentley, M. R., Maltman, A. J. & Fitches, W. R. 1988 Colonsay and Islay: a suspect terrane within the Scottish Caledonides. *Geology* **16**, 26–28.

Bluck, B. J. 1983 Role of the Midland Valley of Scotland in the Caledonian orogeny. *Trans. R. Soc. Edinb. Earth Sci.* **74**, 119–136.

Bluck, B. J. 1985 The Scottish paratectonic Caledonides. *Scott. J. Geol.* **21**, 437–464.

Boccaletti, M., Papani, G., Galeti, R., Rodrigues-Fernandez, J., Lopez Garrido, A. C. & Sanz De Galdeano, C. 1987 Brittle deformation analysis in neotectonics. *Acta Naturalia l'ateneo Parmense* **23**, 179–200.

Crowell, J. C. 1982 The violin breccia, Ridge Basin. *Geological history of the Ridge Basin, Southern California* (ed. J. C. Crowell & M. H. Link), pp. 89–98. Society of Paleontologists and Mineralogists, Pacific Section.

Curry, G. B., Ingham, J. K., Bluck, B. J. & Williams, A. 1982 The significance of a reliable Ordovician age for some Highland Border rocks in Central Scotland. *J. geol. Soc. Lond.* **139**, 451–454.

Cuthbert, S. J., Harvey, M. A. & Carswell, D. 1983 A tectonic model for the metamorphic evolution of the Basal Gneiss Complex, Western South Norway. *J. metamorphic Petrol.* **1**, 63–90.

Dempster, T. J. 1985 Uplift patterns and orogenic evolution in the Scottish Dalradian. *J. geol. Soc. Lond.* **142**, 111–128.

Dempster, T. J. & Bluck, B. J. 1989 The age and origin of boulders in the Highland Border Complex: constraints on terrane movements. *J. geol. Soc. Lond.* **146**, 377–379.

Floyd, J. D. 1982 Stratigraphy of a flysch succession: the Ordovician of West Nithsdale, S.W. Scotland. *Trans. R. Soc. Edinb. Earth Sci.* **73**, 1–9.

Gee, D. 1978 Nappe displacement in the Scandinavian Caledonides. *Tectonophysics* **47**, 393–419.

Harper, D. A. T., Williams, D. M. & Armstrong, H. A. 1989 Stratigraphical correlations adjacent to the Highland Boundary fault in the west of Ireland. *J. geol. Soc. Lond.* **146**, 381–384.

Hartley, J. J. 1933 The geology of north-eastern Tyrone and the adjacent portions of county Londonderry. *Proc. R. Irish Acad.* **B41**, 217–285.

Kelley, S. & Bluck, B. J. 1989 Detrital mineral ages from the Southern Uplands using ^{40}Ar–^{39}Ar laser probe. *J. geol. Soc. Lond.* **146**, 401–403.

Marcantinio, F., Dickin, A. P., McNutt, R. H. & Heaman, L. M. 1988 A 18000-million year old Proterozoic gneiss terrane in Islay with implications for the crustal structure and evolution Britain. *Nature, Lond.* **335**, 62–64.

McCabe, P. J. 1977 Deep distributary channels and giant bedforms in the Upper Carboniferous of the Central Pennines, Northern England. *Sedimentology* **24**, 271–290.

McKerrow, W. S. 1987 Introduction: the Southern Uplands controversy. *J. geol. Soc. Lond.* **144**, 735–736.

Patchett, P. J. & Arndt, N. T. 1986 Nd isotopes and tectonics of the 1.9–1.7 Ga crustal genesis. *Earth planet. Sci. Lett.* **78**, 329–338.

Rogers, G., Dempster, T. J., Bluck, B. J. & Tanner, P. W. G. 1989 A high precision U–Pb age for the Ben Vuirich granite: implications for the evolution of the Scottish Dalradian supergroup. *J. geol. Soc. Lond.* **146**, 789–798.

Allochthonous terranes in the Tethyan Middle East: Anatolia and the surrounding regions

By Y. Yilmaz

İ.T.Ü. Maden Fakültesi, Jeoloji Bölümü, Ayazağa, 80626 Istanbul, Turkey

Turkey consists of a number of tectono-stratigraphic entities corresponding with specific former plate tectonic environments such as island arcs, submarine continental platforms with Atlantic-type and Pacific-type margins. These were successively accreted to Eurasia since the late Palaeozoic resulting in N to S development of the Turkish orogen.

Tethyan evolution of the Anatolia and surrounding regions is the result of demise of the oceanic environments that correspond to the Palaeo-Tethys and Neo-Tethys oceans. The former was totally consumed by Dogger while the latter opened during the Trias-Lias interval and survived into the Cenozoic.

Introduction

The Tethyan evolution of Anatolia is reviewed in terms of two main phases of activity which partly overlap in time: the Palaeo-Tethyan and the Neo-Tethyan. The Palaeo-Tethyan events occurred mainly during the Permian-Lias interval and were most prominent in Northern Anatolia. On the other hand, the Neo-Tethyan phase of activity affected almost the entire Anatolian region from the Triassic through to the Miocene, and its effects linger on to the present day.

The Turkish orogen may be divided into a number of E–W trending belts which are separated from each other by ophiolitic sutures (figure 1). The ages for suturing (Şengör & Yılmaz 1981) suggest that the Anatolian orogen evolved progressively through time from the northern regions to the south.

The purpose of this paper is to review the Mesozoic–Cenozoic tectonic evolution of the Anatolian orogenic belt by emphasizing the geological history of three regions, complementary in nature, which are critical in the evaluation of the consecutive stages of the orogeny through time. These are (1) the Sakarya continent in the west, (2) the Central Pontides in the north, and (3) the southeast Anatolian region in the south (figure 1). These areas have been selected for two main reasons. They have been studied in considerable detail in recent years (Yılmaz 1981; Yılmaz et al. 1984, 1987a; Tüysüz 1990) and they are located in regions where both the Palaeo- and the Neo-Tethyan systems are well exposed and can be reconstructed with some confidence. In the central Pontides the Palaeo–Tethyan and the Neo-Tethyan ophiolitic fragments are observed. In the Sakarya continent remnants of the Karakaya marginal sea and Neo-Tethyan ophiolitic fragments are exposed. In southeast Anatolia Neo-Tethyan ophiolitic fragments, metamorphic massifs, and the later stages in the orogenic evolution can all be evaluated (figure 1).

The Sakarya continent

The sequence in the Sakarya continent is shown in figures 2 and 3. Two different sectors may be distinguished within the region. In the north and the south metamorphic rocks and

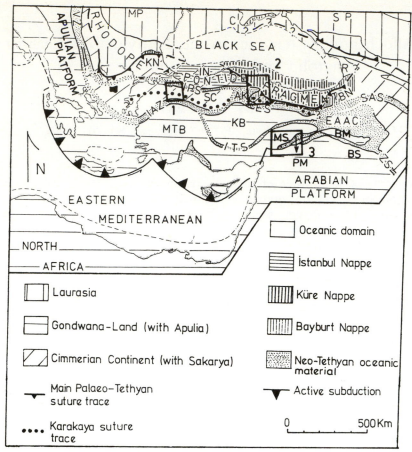

FIGURE 1. Tectonic map showing the Tethyan palaeotectonic elements in Anatolia and the surrounding regions. (Modified after Şengör *et al.* 1984). SP, Schythian platform; C, Crimea; MP, Moesian platform; Sc, Sakarya continent; MTB, Menderes-Taurus block; KB, Kirşehir block (MTB+KB, Anatolide-Tauride platform); PM, Pötürge massif; BM, Bitlis Massif; Vz, Vardar zone; IPS, Intra-Pontide suture; IAZ, Izmir-Ankara suture; AK, Ankara knot; ES, Erzincan suture; ITS, Intra-Tauride suture; SAS, Sevan-Akera suture; EAAC, East Anatolian accretionary complex; BS, Bitlis suture; MS, Maden suture; ZS, Zagros suture; KN, IN and BN are the Kirklareli, Istanbul and Bayburt nappes respectively; R, Riou depression. 1, 2 and 3 are the regions that are described in detail in the text under the headings of the Sakarya Continent, the Central Pontides, and southeast Anatolia respectively.

sedimentary rocks dominate, respectively. Between the two a metamorphosed ophiolitic assemblage and associated *mélange* occur. In the sediment-dominated southern sector there are also some metamorphic rocks occurring at the base of the sequence (figure 3), but these do not resemble the metamorphic rocks of the northern sector in terms of lithology, age, or metamorphic facies.

The southern sector

Two distinct tectonostratigraphic units may be distinguished in the metamorphic rocks of the southern sector (figure 3). One of these is essentially composed of metapelitic rocks that range from a slate-phyllite association to schists and migmatitic gneisses. These are intruded by a post-tectonic granitic pluton (the Söğüt granite) dated isotopically at *ca.* 295 Ma (Çoğulu *et al.* 1965). This association is the oldest rock group of the region and is unconformably overlain by Permian sediments. The other metamorphic unit of the southern sector consists primarily of basic igneous rocks of Triassic age, with associated deep-sea sediments, which

156

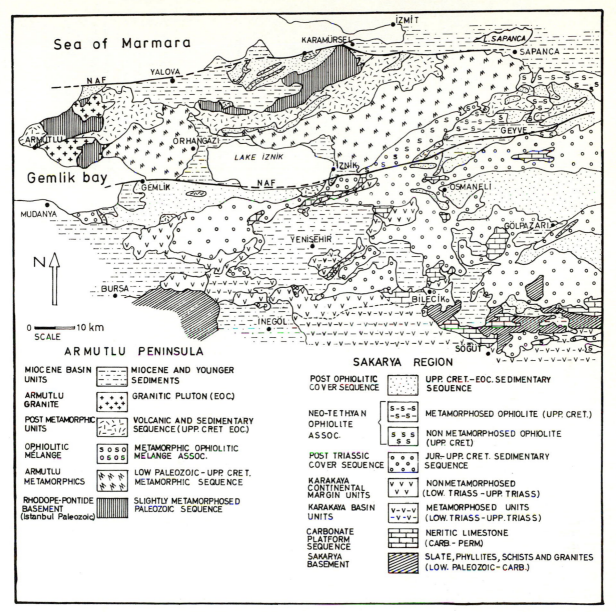

FIGURE 2. Geological map showing different tectono-stratigraphic units of the Sakarya Continent (the region that is shown as no. 1 in figure 1). NAF, branches of the North Anatolian Transform fault.

underwent high-pressure metamorphism. The Pre-Permian metamorphics overlie the Triassic metamorphics and the boundary between the two is a sharp mylonitized thrust contact.

The metamorphic rock units and their tectonic contact are overlain by a transgressive sedimentary sequence which begins with basal sandstones of Liassic age (figure 3). The following succession, as shown in figures 2 and 3 covers a period ranging from the late Jurassic to the Palaeocene.

The northern sector

The metamorphic assemblage of the northern sector forms a sequence (figure 3) more than 1000 m thick in which, despite the greenschist facies metamorphism, well-determined lithological units may be distinguished. In terms of its lithologies and faunas, the sequence may

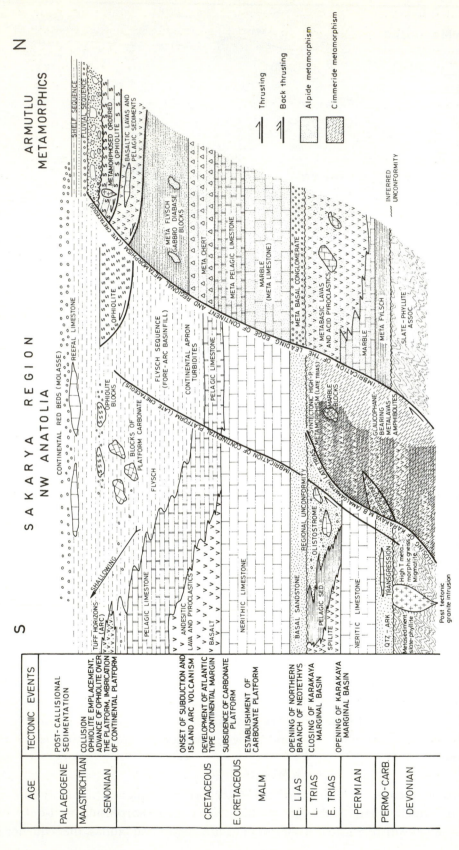

FIGURE 3. Generalized stratigraphic sections across the Sakarya Continent in N–S direction showing the major tectonic elements and tectonic events in time-space reference. The Cimmeride (Triassic) and Alpide (late Cretaceous) stages of metamorphisms are indicated by different symbols that are super-imposed on the lithological symbols.

be closely correlated with the non-metamorphic sequence of the southern sector (figure 3). The metamorphic units appear to represent the lateral, but more basinal, equivalents of the southern sedimentary sequence. Together they were deposited in a continental shelf and slope environment. The oldest sediments that were deposited on top of the northern sector metamorphics and the neighbouring meta-ophiolites are Maastichtian in age, placing a late Cretaceous time limit on the age of metamorphism.

The geological data indicate that the first important tectonic event in the Sakarya continent occurred during the Triassic (figure 3). At the end of the Permian, a carbonate platform was ruptured leading to the opening of a short-lived basin, known as the Karakaya marginal sea (Şengör & Yılmaz 1981). Its remnants can now be traced along the Pontides, in the Taurus mountains, and within the southeast Anatolian metamorphic massifs (figure 1) extending into Iran (Yılmaz *et al.* 1987*b*). Accompanying this event, a typical ocean floor, represented by an almost complete ophiolite sequence, was formed. At the end of the Triassic the basin was closed (figure 3), resulting in the northwards thrusting of the allochthons. The ophiolites were metamorphosed to glaucophane-bearing greenschists facies.

Following the late Triassic closure of the Karakaya marginal basin a new rifting phase began during the Liassic (Yılmaz 1981; Şengör & Yılmaz 1981). The rift-related sequence formed at that time can be traced throughout the Pontides (Görür *et al.* 1983) as far as Iran. During the Jurassic and early Cretaceous the region was covered by platform carbonates (figure 3). This platform was bordered by the oceanic environments in the north and south (Şengör & Yılmaz 1981). During the early late Cretaceous, platform subsidence occurred, as is evidenced by deposition of the pelagic limestones and cherts (figure 3). In the late Cretaceous, not long after the regional subsidence of the continental platform, a new phase of tectonism was initiated. The onset of tectonism is shown by the flysch deposition over units of variable age, thrusts, transportation of blocks of the older rock units into the flysch trough, and the influx of ophiolitic detritus (figure 3). All of these associated events may be regarded as linked with the emplacement of the ophiolite onto the continental platform (figure 3). At the same time the leading edge of the Sakarya continent was initially internally imbricated (figure 3), and was then buried and metamorphosed together with an obducted slab of the ophiolite to form the metamorphic assemblage of the northern sector.

These related events correspond to the demise of the northern branch of the Neo-Tethyan ocean floor and thus to the continental collision which occurred between the Sakarya continent and the Pontides during the late Cretaceous (Şengör & Yılmaz 1981).

The central Pontides

The central Pontides form a geological mosaic (figure 4) that is composed of amalgamation of various, roughly E–W trending accreted tectonstratigraphic units (figures 1 and 4). Among them the Ankara–Erzincan ophiolitic suture, the Karakaya suture, the eastern Pontide units (the Bayburt nappe), the western Pontide units (the Istanbul nappe) and the eastern extension of the Sakarya continent may be mentioned (Tüysüz 1990) (figure 1).

Figure 5 shows a N–S geological section across the central Pontides. The section shows that in the south central part of the region Palaeozoic metamorphosed rocks occur at the base of the sequence. Onto this, different units, of ophiolitic rocks and associated assemblages, were emplaced initially from the north during the Middle Mesozoic, and then later from the south at the end of the Mesozoic. The older ophiolites, which outcrop in the northern part of the

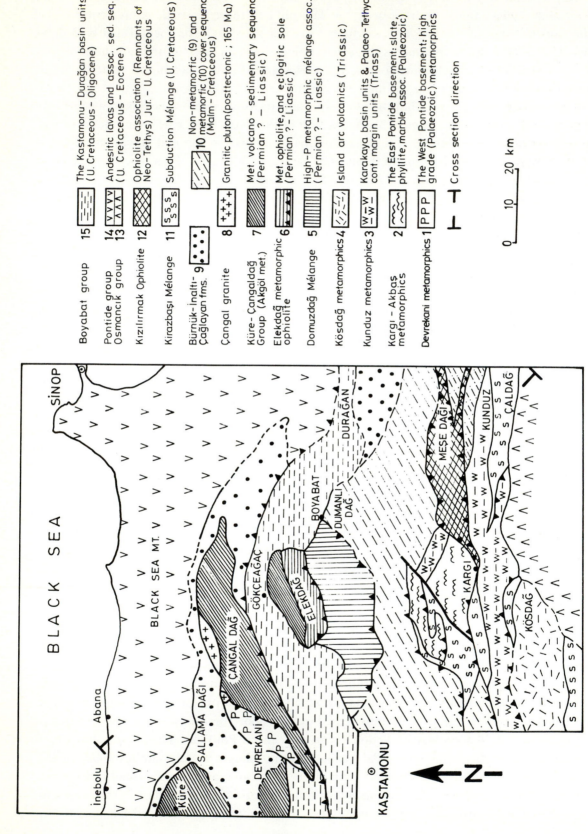

FIGURE 4. Geological map showing different tectono-stratigraphic units of the Central Pontides (the region that is shown as no. 2 in figure 1). The cross-section direction indicates the section shown in the figure 5, 9 and 10 are the identical units of which 9 is metamorphic, 10 is non-metamorphic.

Legend:

15 Boyabat group — The Kastamonu–Durağan basin units (U. Cretaceous – Oligocene)

14 Pontide group
13 Osmancık group — Andesitic lavas and assoc. sed. seq. (U. Cretaceous – Eocene)

12 Kızılırmak Ophiolite — Ophiolite association (Remnants of Neo-Tethys) Jur. – U. Cretaceous

11 Kirazbaşı Mélange — Subduction Mélange (U. Cretaceous)

9 Bürnük-İnaltı-Çağlayan fms. — Non-metamorfic (9) and 10 metamorfic (10) cover sequence (Malm – Cretaceous)

8 Çangal granite — Granitic pluton (posttectonic ; 165 Ma)

7 Küre–Çangaldağ Group (Akgöl met.) — Met. volcano – sedimentary sequence (Permian ? – Liassic)

6 Elekdağ metamorphic ophiolite — Met. ophiolite, and eclogitic sole (Permian ? – Liassic)

5 Domuzdağ Mélange — High-P metamorphic Mélange assoc. (Permian ? – Liassic)

4 Kösdağ metamorphics — Island arc volcanics (Triassic)

3 Kunduz metamorphics — Karakaya basin units & Palaeo-Tethyan cont. margin units (Triass)

2 Kargı – Akbaş metamorphics — The East Pontide basement: slate, phyllite, marble assoc. (Palaeozoic)

1 Devrekani metamorphics — The West Pontide basement: high grade (Palaeozoic) metamorphics

Cross section direction

0   10   20 km

region, are metamorphosed and are regarded to be remnants of the Palaeo-Tethyan ocean (Şengör *et al.* 1984).

The upper layers of these ophiolites, including the pillow lavas, occur in the Küre and Çangal Dağ areas (figure 4) where the epiophiolitic deep-sea sediments yielded ages ranging from Permian (?) to Liassic (Aydın *et al.* 1986). The lower layers outcrop in the Elekdağ area, where a thin eclogitic sole is seen to be attached to the ophiolitic base (figure 5). Together with this sliver of eclogite the ophiolites were thrust over a mélange association. The ophiolite was then emplaced southwards possibly as a backthrust onto the continental margin sequence as a nappe package.

At the base of the continental margin sequence there is a slate-phyllite association of possible lower Palaeozoic age. This is overlain by metamorphosed Permian carbonates followed by a basaltic metalava sequence that contains blocks of the underlying carbonates. The metalavas are intercalated with metapelitic rock units which include blocks of gabbro. Towards the upper levels of the sequence, metamorphosed pelagic sediments, intercalate with an upper unit of basaltic metalavas. At the top of the sequence metamorphosed olistostromes containing abundant ophiolitic fragments occur (figure 5). The units overlying the metamorphosed Permian carbonate succession are identical to the Triassic of the Sakarya continent.

The metamorphics are cut by a post-tectonic granitic intrusion (figure 5) which has been dated isotopically at 165 Ma (Yılmaz 1979). The granite, along with the underlying units, is overlain by a first common regional sedimentary cover sequence of Malm age (fiugre 5). However, post-deformation sedimentation commenced, locally earlier than this, in the Liassic. Deposition at first occurred only on the metamorphosed continental margin sequence, but it continued during the Malm. Sediment deposition was uninterrupted in areas which are unaffected by the Palaeo-Tethyan ophiolite obduction.

The region-wide transgressive Malm sequence (figure 5) began with basal sandstones and conglomerates. These were followed by neritic carbonates, which change laterally to shales, of Lower Cretaceous age. The Upper Cretaceous is represented by a flysch succession. In the northern parts of the region (the Central Black Sea Mountains) this is post-dated by calc-alkaline andesitic volcanism. In the south it is initially overlain by a wildflysch and then a flysch deposition of an ophiolitic nature, and these are then followed by the emplacement of a thick ophiolite nappe package consisting of slices of an internally ordered ophiolite sequence. In contrast to the northern ophiolites (the Palaeo-Tethyan ophiolites) that were emplaced during the Dogger, the southern ophiolites are not metamorphosed.

The geological evolution of the central Pontides shows some important similarities to the Sakarya region, especially where the Karakaya marginal basin events are recorded. The data indicate that the Karakaya basin closed at the end of the Triassic at a time when the Palaeo-Tethyan ocean was still present in the north. The demise of Palaeo-Tethys and the obduction of ophiolitic slabs took place in the Dogger (figure 5). Immediately after the closure of the Karakaya marginal basin a new rifting event started in the southern sector of the Pontides (Yılmaz 1981; Şengör & Yılmaz 1981) during the Liassic, and this eventually led to the opening of the Neo-Tethyan ocean in the southern part of the Pontides.

During the Malm, the Palaeo-Tethyan events ended completely, and a south-facing Atlantic-type continental margin became established by the beginning of the early Cretaceous. During the late Cretaceous, the Neo-Tethyan ocean began to be subducted along a north-dipping Benioff zone under the ophiolite-laden northern continent (Şengör & Yılmaz 1981).

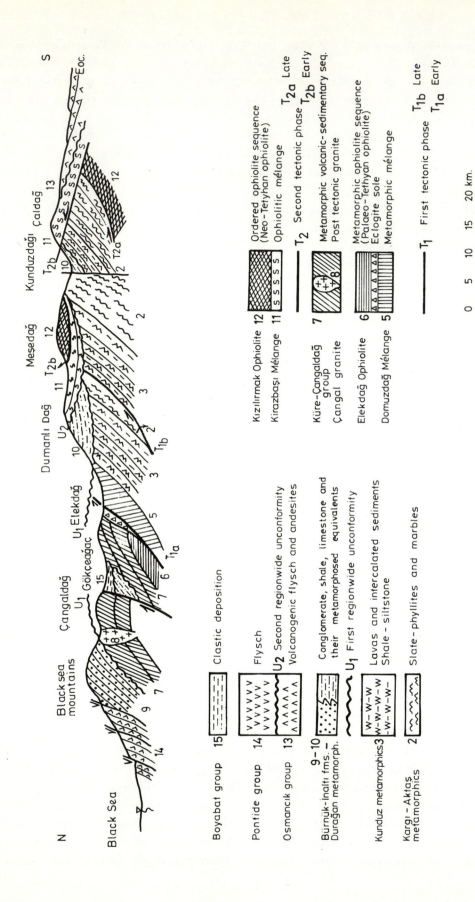

Figure 5. Geological section across the Central Pontides. Numbers refer to the rock units shown in figure 4. Abbreviations: Dağ, Mt, Mountain.

TERRANES IN THE TETHYAN MIDDLE EAST

Subduction generated an Andean-type magmatic arc on the Upper Cretaceous units in the northern parts of the region, while in the south a subduction *mélange* was produced in front of the continent. At the end of the Cretaceous ophiolite slabs were obducted onto the continental margin sequence along the southern edge of the continent (figure 5), possibly by a *retrocharriage* mechanism (Yılmaz & Tüysüz 1984; Tüysüz 1990).

THE SOUTHEAST ANATOLIA

In the region of southeast Anatolia 3 geologically different zones may be distinguished. From south to north these are (1) the Arabian platform, (2) the zone of imbrication, and (3) the nappe region (figures 6 and 7).

The Arabian platform

The Arabian platform represents the Arabian authochtonous and parauthochtonous sedimentary succession (figures 6, 7 and 8) that has accumulated, almost without interruption, since the early Palaeozoic on a stabilized craton (the Pan-African basement). The platform sequence is therefore relatively undeformed in its exterior parts. However, towards the north, where the region was affected by the southeast Anatolian orogen, the platform sequence gradually becomes more deformed into a foreland fold and thrust belt.

The Mesozoic sequence, and the associated geological evolution of southeast Anatolia until the ophiolite obduction onto the Arabian platform in the late Cretaceous appear similar in many ways to the other peripheral parts of the Arabian platform (see, for example, Robertson 1987).

However, the post-ophiolite geology of the Arabian platform in southeast Anatolia differs significantly from the other peri-Arabian regions as well as from the other regions of Anatolia. In this region, contrary to the other areas, the orogenic evolution had not ended with ophiolite obduction in the late Cretaceous, but extended into the late Cenozoic.

Figure 8 shows the post-ophiolite stratigraphy in the north of the Arabian platform and farther north where a marine environment was re-established over the continental platform at the end of Mesozoic (late Maastrichtian), and it survived until the end of the early Miocene with only one major phase of non-deposition during the late early Eocene (figure 8). Towards the north this platform joined an oceanic environment during the early–middle Tertiary as evidenced by co-eval units that occur in the zone of imbrication and along the Missis-Andırın Mountains (figures 6 and 8).

The zone of imbrication

The zone of imbrication is a narrow E–W trending belt which is seen to have been squeezed and sandwiched between the Arabian platform to the south and the nappe regions to the north (figures 6 and 7). It is separated from the other two regions by thrusts. The zone consists of a number of imbricated thrust slices which cover a succession comprising the late Cretaceous–early Miocene period. Between the thrust slices the stratigraphic sequences are reversed with older units overlying younger units (figures 7 and 8). The base of the section is represented by a flysch succession which appears to be laterally equivalent to the similar flysch at the top of the Arabian platform sequence. Although contacts between the units, described above, are tectonic in nature (thrusts), there is clear evidence within each slice to suggest that the slices are genetically related units which formed one continuous succession before imbrication.

Towards the west, the zone of imbrication extends to the Missis-Andırın Mountain range

163

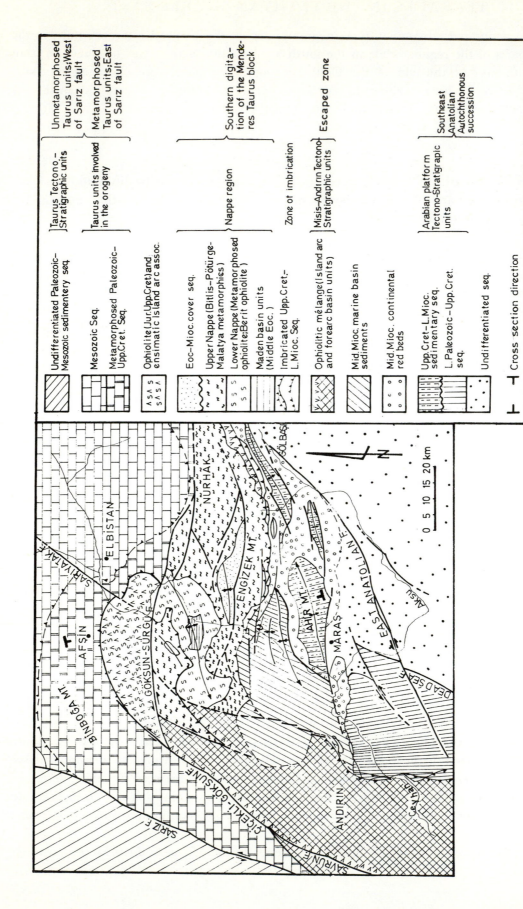

FIGURE 6. Geological map showing different tectono-stratigraphic units of the south-east Anatolia in the Maraş-Elbistan regions (the region that is shown as no. 3 in figure 1). The cross-section direction indicates the section shown in the figure 7.

Legend:

Taurus Tectono-Stratigraphic units
- Taurus units involved in the orogeny
  - Undifferentiated Paleozoic-Mesozoic sedimentery seq. — Unmetamorphosed Taurus units; West of Sariz fault
  - Mesozoic Seq.
  - Metamorphosed Paleozoic-Upp.Cret. Seq. — Metamorphosed Taurus units; East of Sariz fault
  - Ophiolite(Jur.Upp.Cret.)and ensimatic island arc assoc.

Nappe region
- Eoc-Mioc. cover seq. — Southern digitation of the Menderes Taurus block
- Upper Nappe (Bitlis-Pötürge-Malatya metamorphics)
- Lower Nappe (Metamorphosed ophiolite;Berit ophiolite)
- Maden basin units (Middle Eoc.)
- Imbricated Upp.Cret.-L.Mioc. Seq. — Zone of imbrication

Misis-Andırın Tectono-Stratigraphic units
- Ophiolitic mélange(Island arc and forearc basin units) — Escaped zone
- Mid.Mioc. marine basin sediments
- Mid.Mioc. continental red beds

Arabian platform Tectono-Stratigrapic units
- Upp.Cret-L.Mioc. sedimentary seq. — Southeast Anatolian Autochthonous succession
- L.Paleozoic-Upp.Cret. seq.
- Undifferentiated seq.

Cross section direction

0 5 10 15 20 km

Labels on map: SARIVAYAK, BİNBOĞA MT., ELBİSTAN, AFŞİN, NURHAK, GÖKSUN-SÜRGÜ F., ENGİZEK MT., SÖLBAŞI, AHIR MT., MARAŞ, EAST ANATOLIAN F., AKSU, DEAD SEA F., ANDIRIN, ÇİÇEKLİ-GÖKSUN F., SAVRUN F., SARIZ F., Ceyhan, N

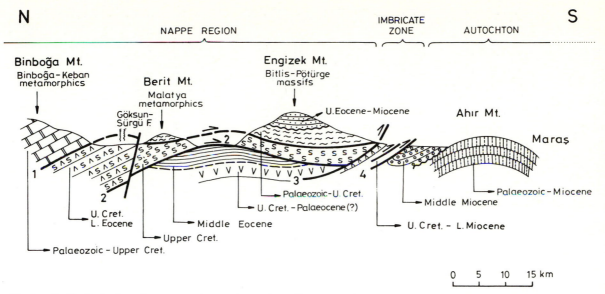

FIGURE 7. Geological section across the southeast Anatolian orogenic belt between Maraş and Afşin (see the map in figure 6). Numbers 1 to 4 indicate progressive southerly movement of the nappes. Main thrusting stages: 1, late early Ecocene; 2, late middle Ecocene; 3, early Miocene; 4, late early Miocene–middle Miocene. They correspond to numbers 2 to 5 respectively in Figure 8.

(figure 6), which forms a belt a few tens of kilometres wide. The rock units occurring along this mountain range are exposed as a 100 m belt within the imbricated zone. This observation suggests that the Missis-Andırın Mountains represent a region that has not been shortened to the same degree between the converging plates that formed the southeast Anatolian orogen. Therefore the Missis-Andırın Mountains may possibly be an escaped zone (figure 6). In fact the equivalent of the mélange units which form the basement of the Missis-Andırın belt is only barely present along the zone of imbrication where they are thought to have been totally subducted.

The Maastrichtian–Lower Eocene sedimentary units of the zone of imbrication appear to be the lateral, but more distal, equivalents of the Arabian platform units, distal either towards the northerly situated continental slope and the abyssal plain (Yılmaz *et al.* 1987*a*).

The Nappe region

To the north of the zone of imbrication is the Nappe region (figures 6, 7 and 8). The nappes are the structurally highest tectonic units of the southeast Anatolian orogenic belt (figures 7 and 8). They are composed of two large nappe packages: the lower nappes and the upper nappes. The lower nappe package consists primarily of ophiolitic rocks. In contrast, the upper nappe package is formed from the metamorphic massifs of southeast Anatolia, i.e. the Bitlis and Pötürge massifs (figures 1 and 7) or the Malatya, Keban and Engizek metamorphics. The upper nappe package rests on a gently folded thrust surface on top of the lower nappe package (figure 7).

The geological data indicate that the upper nappe package initially moved over the lower package at the end of the early Eocene after which the two packages were transported collectively (figures 7 and 8). The ophiolite nappe contains all of the layers of an ordered ophiolite sequence which has undergone polyphase metamorphism (Yılmaz *et al.* 1987*a*).

The upper nappe package is dominantly comprised of a regionally metamorphosed

165

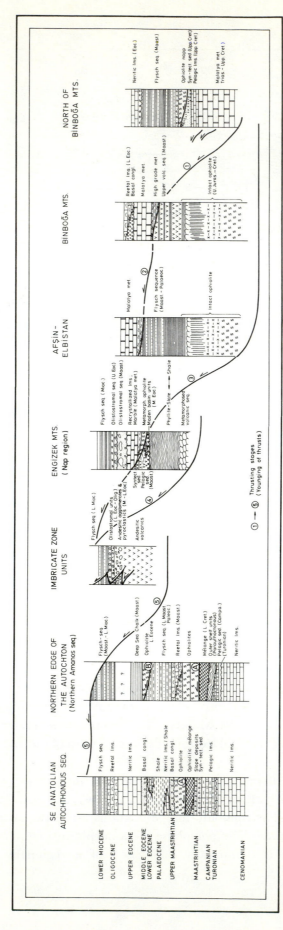

FIGURE 8. Generalized stratigraphic sections from the different regions of the south-east Anatolia across the orogenic belt (see the map in figure 6 for the locations). Thrusting stages indicate progressive southerly movement of the nappes towards the Arabian platform during the late Cretaceous–middle Miocene period. Main thrusting stages are; 1, late Cretaceous (early Maastrichtian); 2, early Eocene; 3, late middle Eocene; 4, early Miocene; 5, late early Miocene–middle Miocene. Abbreviations (in bracket); L, lower; M, middle; U, Upper; MS, limestone; Congl, Conglomerate; Agl, Agglomerate; Syn-tect sed, Syntectonic sediment deposition; slate shale, shale–slate transition; seq, sequence; met, metamorphic units. A and B indicate ophiolite emplacement on to the Arabian platform during the late Cretaceous and middle Eocene respectively. Repetitive arrows indicate imbrication.

sedimentary sequence which ranges in age from the Palaeozoic to the Upper Cretaceous. Within this sequence two parts may be clearly distinguished: (a) high-grade metamorphic schists and gneisses as core rocks, (b) low-grade metamorphic slate-phyllites and marbles as the envelope rocks. The oldest sedimentary rocks to be deposited on top of the metamorphic rocks of the upper nappe are clastic sediments of Upper Maastrichtian age. The age of metamorphism may therefore be well constrained as late Cretaceous, most probably late Campanian–early Maastrichtian, because the youngest metamorphic rocks are Campanian pelagic limestones (Yılmaz *et al.* 1987 *a*).

Collectively the data (figure 8) suggest that the cause of the orogeny was the progressive elimination and eventual demise of an ocean which existed in the northern part of the Arabian plate. During closure of the ocean, convergence initially led to formation of a collision between an island arc and a continent (between the Yüksekova complex and the Taurides) at the end of early Eocene. This deformed package then collided with the Arabian platform when the remainder of the ocean had been entirely consumed during late middle Eocene–late Eocene (Yılmaz *et al.* 1987 *a*). Thus the southeast Anatolian orogenic evolution may be summarized as the progressive relative (southerly) movement of the northerly situated allochthons (the nappes) towards the Arabian platform during the late Cretaceous–early Miocene interval (figure 8). During these events, new tectonic units (an island arc association: the Helete Volcanics and the Yüksekova complex); a back arc basin unit: the Maden Complex; a remnant sea infill: the Çüngüş Complex; and a syntectonic flysch deposition: the Lice Fm.) were progressively amalgamated. The allochthons, as a package, were thrust over the authochthonous sequence at a later stage in the orogeny, during the late early Miocene. The zone of imbrication which lies adjacent to the nappes may thus be regarded as a region which was bulldozed in front of the southwards moving nappe pile.

Post-collisional convergence after the middle Miocene began to be largely accommodated by E–W trending strike–slip faults leading to the reactivation of earlier thrusts and to their dissection by high-angle faults.

Conclusion

In the above paragraphs I reviewed, in as much detail as allowed by the available space, the tectonic evolution of three complicated regions from the Turkish orogenic collage. Throughout these paragraphs I placed interpretative labels on the tectono-stratigraphic units I defined. Nowhere in the analysis presented above did I follow the recommended 'terrane analysis' (see, for example, Jones *et al.* 1983). Before the advent of plate tectonics, the Turkish geology had reached a cul-de-sac. We had mapped many regions at a scale of 1:25000 and we had stopped dead in front of insurmountable difficulties of not being able to make sense out of the fragmentary and vastly complicated record. Şengör & Yılmaz (1981) was our first venture into the interpretation of the vast database in terms of plate tectonics. That paper suddenly lifted the mist and the progress since then in Turkish geology has been immense. Therefore I can see no advantage in a return to this tectonic taxonomy.

I thank Dr A. M. C. Şengör and Dr D. Latin for improving the text and giving useful comments.

REFERENCES

Aydın, M., Şahintürk, Ö., Serdar, H. S. & Özçelik, Y. 1986 *Türkiye Jeoloji Kurumu Bülteni* **29** (2), 1–16.
Çoğulu, E., Delaloye, M. & Chessex, R. 1965 *Arch. Sci. Genéve* **18** (3), 692–699.
Görür, N., Sengör, A. M. C., Yılmaz, Y. & Akkök, R. 1983 *Türkiye Jeoloji Kurumu Bülteni* **26**, 11–20.
Jones, D. L., Howell, D. G., Coney, P. J. & Monger, J. W. H. 1983 *Accretion tectonics in the Circum-Pacific regions* (ed. M. Hashimotu & S. Uyeda), pp. 21–35. Tokyo: Terrapub.
Robertson, A. 1987 *Bull. Geol. Soc. Am.* **99**, 633–653.
Şengör, A. M. C. 1979 *Nature, Lond.* **279**, 590–593.
Şengör, A. M. C. & Yılmaz, Y. 1981 *Tectonophysics* **75**, 181–241.
Şengör, A. M. C., Yılmaz, Y. & Sungurlu, O. 1984 *Geol. Soc. Lond., Spec. Rep.* **14**, 117–152.
Tüysüz, O. 1990 *Tectonic evolution of a part of the Tethyside orogenic collage: The Kargı Massif, northern Turkey.* Tectonics.
Yılmaz, O. 1979 *Daday-Devrekani Masifi Kuzeydoğu Kesimi Metamorfitleri.* Doçentlik Tezi, Hacettepe Üniversitesi. (243 pages.)
Yılmaz, Y. 1981 *İstanbul Yerbilimleri* **1** (1–2), 33–52.
Yılmaz, Y., Gürpınar, O., Kozlu, H., Gül, M. A., Yiğitbaş, Yıldırım, M., Genç, C. & Keskin, M. 1987*a Maraş Kuzeyinin Jeoloisi (Andırın-Berit-Engizek-Nurhak-Binboğa Dağları).* Türkiye Petrolleri, A. O. Rep. no. 2028. (218 pages.)
Yılmaz, Y., Yiğitbaş, E. & Yıldırım, M. 1987*b 7th Petrol. Cong. Proc. Ankara*, pp. 65–78.

Discussion

P. D. CLIFT (*Grant Institute, University of Edinburgh, U.K.*). The proposed model for the tectonic evolution of Anatolia and the Tethyan Middle East envisages a progressive accretion of continental terranes to the active northern margin of the Neotethys, following rifting from Gonwana and a relatively straightforward south-to-north migration. This is in marked contrast to models of other orogenic belts, e.g. the North American Cordillera and the North Atlantic Caledonides. The structural evidence so far gathered indicates a simple strike–perpendicular thrust shortening, with little indication of large degrees of motion of accreted terranes along the active margin. The relatively small width of the Neotethys (*ca.* 100–1000 km) in this region makes the use of faunal provinces for tracking block motions impossible. In addition the east–west trend of the belt means palaeomagnetism cannot be used to follow margin parallel displacements. Some quantification of the amounts of along strike motion and potential sites of the 'home ports' of these continental terranes may be obtainable by application of isotopic ($^{87/86}$Sr or Sm/Nd T_{CR}) provenance studies to these regions. Pursuit of these methods, which has yielded so much information in the Caledonides may help us to be more certain whether the Neotethys really does provide us with an important exception to the Pacific-type terrane accretion model or not.

Y. YILMAZ. In its general frame I agree with Dr Clift's comment. However, the possibility of margin-parallel displacement in the Anatolian orogen should not completely be ruled out. I suspect it especially in the eastern Anatolia from facies arguments and, timing and nature of tectonic events of Trias–Lias and late Cretaceous–Eocene periods (see Yılmaz *et al.* 1987*a, b*), possibly similar to that as shown by A. M. C. Şengör on his recent accounts about Iran. In this paper I only presented data within the limit of the space allowed to me. Therefore I could not enter more detailed discussion on this topic.

It is my view, however, that the large, orogen-parallel strike–slip can be detected by field geological methods as long as detailed regional and relevant data are available. More sophisticated approaches of the kind Dr Clift points out can be used more effectively as additional supporting data in the regions which underwent multistages of orogenic deformation where earlier events were obscured by later events.

Allochthonous terrane processes in Southeast Asia

By I. Metcalfe†

Institute for Advanced Studies, University of Malaya, 59100 Kuala Lumpur, Malaysia

Southeast Asia comprises a complex agglomeration of allochthonous terranes located at the zone of convergence between the Eurasian, Indo-Australian and Philippine Sea plates. The older continental 'core' comprises four principal terranes, South China, Indochina, Sibumasu and East Malaya, derived from Gondwana-Land and assembled between the Carboniferous and the late Triassic. Other terranes (Mount Victoria Land, Sikuleh, Natal, Semitau and S.W. Borneo) were added to this 'core' during the Jurassic and Cretaceous to form 'Sundaland'.

Eastern Southeast Asia (N. and E. Borneo, the Philippines and eastern Indonesia) comprises fragments rifted from the Australian and South China margins during the late Mesozoic and Cenozoic which, together with subduction complexes, island arcs and marginal seas, form a complex heterogeneous basement now largely covered by Cenozoic sediments. Strike–slip motions and complex rotations, due to subduction and rifting processes and the collisions of India with Eurasia and Australia with Southeast Asia, have further complicated the spatial distribution of these Southeast Asian terranes.

A series of palinspastic maps showing the interpreted rift–drift–amalgamation–accretion history of Southeast Asia are presented.

Introduction

Southeast Asia is a complex composite of allochthonous continental fragments, subduction complexes and small ocean basins located at the zone of convergence between the Eurasian, Indo-Australian and Philippine Sea Plates. The older cratonic core (figure 1) comprises four major continental terranes, South China, Indochina, Sibumasu and East Malaya, bounded by sutures representing former oceans and now recognized along mobile belts by ophiolites, melanges, volcano-plutonic arcs and accompanied by major tectonic faults and lineaments. Details of the rift–drift–amalgamation–accretion history of these four blocks are still contentious due to the lack of sufficient constraining data. It is generally agreed that the major allochthonous terranes of the region had their origin on the northeastern margin of Gondwana-Land in the Southern Hemisphere (Audley-Charles, 1983; Audley-Charles *et al.* 1988; Sengor 1984; Metcalfe 1988; Burrett & Stait 1986, to mention but a few). Major differences of opinion exist, however, regarding the time of rifting of Sibumasu from Gondwana-Land (late Jurassic according to Audley-Charles (1988) and Audley-Charles *et al.* (1988); late Permian according to Sengor *et al.* (1988); late early Permian according to Metcalfe (1988); and late Devonian according to Bunopas (1982) and Burrett & Long (1989). Timings of amalgamation and accretion are equally contentious (e.g. Indochina and South China sutured in the Carboniferous according to Gatinsky *et al.* (1984); in the Late Triassic according to Sengor (1984), Sengor *et al.* (1988); Sibumasu sutured to Indochina and East Malaya in the Permian according to Helmcke (1983, 1985); in the early Triassic according to Mitchell (1989); in the

† Present address: Department of Earth Sciences, University of Oxford, Parks Road, Oxford OX1 3PR, U.K.

169

late Triassic according to Sengor (1984), Sengor *et al.* (1988); or in the middle–late Cretaceous according to Audley-Charles (1988).

A number of continental terranes were added to the cratonic core of Southeast Asia during the Mesozoic, including the Hainan Island terranes on the northeast (Yu 1989), Mount Victoria Land and the Sikuleh and Natal terranes (figure 1) on the (present) western side (Mitchell 1989; Pulunggono & Cameron 1984) and the Semitau and S.W. Borneo terranes on the southeastern side (Metcalfe 1988, this paper; Williams *et al.* 1986) to form 'Sundaland'. The eastern part of Southeast Asia, east of Sundaland and west of the Philippine Trench and extending from Hainan down to Australia (figure 1), comprises a complex assemblage of continental fragments, stretched continental crust, subduction complexes, island arcs and small ocean basins. The continental fragments in this region appear to be derived from the margin of the South China terrane (Holloway 1981; Taylor & Hayes 1983) or from the Papua New Guinea–N. Queensland margins of Australia (Pigram & Panggabean 1984; Audley-Charles *et al.* 1988).

Cenozoic modification of the region has largely been attributed to expansion and the east and southeast extrusion of Southeast Asia along major left-lateral strike–slip faults consequent

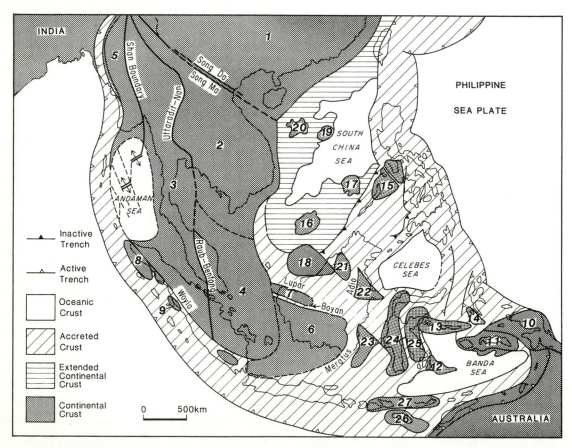

FIGURE 1. Map showing the continental allochthonous terranes and principal sutures of Southeast Asia: 1, South China; 2, Indochina; 3, Sibumasu; 4, East Malaya; 5, Mount Victoria Land; 6, S.W. Borneo; 7, Semitau; 8, Sikuleh; 9, Natal; 10, West Irian Jaya; 11, Buru-Seram; 12, Buton; 13, Bangai-Sula; 14, Obi-Bacan; 15, North Palawan; 16, Spratley Islands–Dangerous Ground; 17, Reed Bank; 18, Luconia; 19, Macclesfield Bank; 20, Paracel Islands; 21, Kelabit-Longbowan; 22, Mangkalihat; 23, Paternoster; 24, West Sulawesi; 25, East Sulawesi; 26, Sumba; 27, Banda Allochthon.

upon the collision of India with Eurasia (Tapponier *et al.* 1982; Tapponier *et al.* 1986). An alternative model for the South China Sea region has been put forward by Taylor & Hayes (1980, 1983) and Holloway (1981, 1982) which involves rifting and associated southwards subduction of Mesozoic oceanic crust beneath Borneo. In this model, the North Palawan, Reed Bank, Dangerous Ground, Spratley Islands and Luconia terranes rift from the South China–Indochina margin and travel south by continental stretching in the west and active spreading of the South China Sea in the east. Counter-clockwise rotations of Borneo and the Philippines are proposed in this model and the importance of right-lateral strike–slip motions emphasized.

THE ALLOCHTHONOUS TERRANES AND THEIR ORIGINS

Terranes here considered truly allochthonous are restricted to those continental blocks and fragments (figure 1) which have travelled significant distances from their sites of origin. The South China superterrane comprises at least two welded blocks. Klimetz (1987) proposed three component terranes, the Yangzi Craton, the Jiangnan Terrane, both with Proterozoic basements, and the Cathaysian Terrane, with a pre-Devonian basement. Sengor (1984, 1989), Sengor & Hsu (1984) and Hsu *et al.* (1988) consider it comprises two terranes, the Sichuan Block and the Southeast China Block (or Yangtze and Huanan blocks according to Sengor *et al.* (1988)) which sutured together in the Late Triassic or Jurassic. Wang (1986), Charvet & Faure (1989) and Ren (1989) dispute this late Mesozoic collision and suggest it was in fact a Proterozoic event! The term South China Block is here used to include both the Yangtze and Huanan blocks for the purposes of discussions in this paper. It appears to have its origin in the Western Himalaya–Iran region of Gondwana-Land from where it rifted in the Silurian or Devonian (Lin *et al.* 1985; J.-L. Lin, this Symposium). Indochina is a long stable Pan-African continental block with a pre-Cambrian basement generally overlain by Palaeozoic shallow marine and Mesozoic continental deposits. Its origin is believed to be on the margin of eastern Gondwana-Land (Audley-Charles 1983, 1988). See Metcalfe (1988a) for details. The Sibumasu terrane (Metcalfe 1986, 1988a) is an elongate continental block with a Proterozoic basement overlain by Palaeozoic shallow marine sequences including late Carboniferous–early Permian glacial-marine diamictites. Mesozoic strata comprise shallow marine and continental deposits with deeper marine sequences developed in certain basins (e.g. Lampang Basin, Semanggol Basin). Sibumasu is regarded to have originated on the northwest margin of Australian Gondwana-Land (Sengor & Hsu 1984; Burrett & Stait 1986; Audley-Charles 1988; Metcalfe 1984, 1986, 1988a). The East Malaya terrane has a Proterozoic basement indicated by zircon inheritance age determinations from granites (Liew & McCulloch 1985), but the oldest exposed rocks are Palaeozoic shallow marine strata, overlain by Triassic marine volcaniclastics near its western margin and by Jurassic and Cretaceous continental rocks to the east.

The pre-Mesozoic 'South West Borneo' block of Metcalfe (1986, 1988) is now known to be composite comprising the Semitau Terrane, a small continental block occurring between the Lubok Antu and Boyan accretionary complexes and the South West Borneo block comprising the rest of S.W. Borneo, south of the Boyan melange belt and west of the Meratus accretionary complex. The Mount Victoria Land terrane has a schist basement of pre-Mesozoic age overlain by Triassic turbidites and Cretaceous (Albian) ammonite-bearing shales and limestones in the Indoburman Ranges and by a late Mesozoic–Cenozoic volcanic arc association in the Central

Lowlands of Burma. It is believed to be derived from the N.W. Australian margin of Gondwana-Land. The Sikuleh terrane has a basement of quartzites and phyllites of probable Palaeozoic age, intruded by calc-alkaline granitoids and Tertiary rhyolites and Mo-bearing breccia pipes (Cameron *et al.* 1980). Its origin remains in doubt (Cameron *et al.* 1980) regarding it as a rifted fragment of Sundaland in a marginal basin setting while Pulunggono (1983) and Barber (1985) prefer a subduction complex model with, presumably, a Gondwana-Land origin for Sikuleh. The related Natal terrane and two other probable continental fragments of smaller size located on the southwest margin of Sumatra (figure 1) are also possibly small fragments of Gondwana-Land. Recent work has shown that Hainan Island includes two small continental fragments, the Qiongzhong and Yaxian terranes, that were derived from N.W. Australian Gondwana-Land (Yu 1989). The Palaeozoic sequences on these small blocks are similar to those of N.W. Australia and yield trilobites and other invertebrates endemic to Australia and Gondwana-Land. The sequence also includes Lower Permian glacial-marine diamictites. The Mangkalihat terrane is poorly known but a continental basement is indicated by tin-bearing granitoids, and early Devonian coral-bearing limestones (Rutten 1940). Rocks such as andesite, dacite, radiolarian chert and limestone suggest an old island arc association (Hutchison 1990). The origin of this terrane is unknown but thought likely to be similar to that of the Semitau terrane.

The other pre-Mesozoic continental terranes of the region comprise a number of small continental blocks of either South China or Australian affinities and origin. The terranes showing affinities to South China include the Paracel Islands terrane which has a Palaeozoic and possibly Pre-Cambrian basement with a thin sedimentary cover; the poorly known Macclesfield Bank terrane with a presumed Palaeozoic or older basement; the Spratley Islands–Dangerous Ground–Reed Bank terranes which have a presumed Palaeozoic basement overlain by slightly metamorphosed deltaic sandstones and siltstones and dark green mudstones with plants and bivalves of Triassic and Jurassic age; the North Palawan Block which has a late Palaeozoic basement of Middle Permian fusulinid limestones and Middle Triassic cherts in North Palawan, and a metamorphic complex of mica schist, slate, quartzite and marble in Mindoro; and the Luconia and Kelabit-Longbowan terranes which are stable areas with shallow marine Tertiary reef limestones or fluvio-deltaic deposits surrounded by pene-contemporaneous deeper-water sediments. The Luconia area also exhibits high geothermal gradients and the Kelabit-Longbowan area has numerous salt springs indicating subsurface salt deposits (Hutchison 1989).

The pre-Mesozoic terranes derived from the Australian margin include West Irian Jaya (Pigram & Panggabean 1984), with a basement of Palaeozoic metasedimentary rocks and an origin in the Papua New Guinea or N. Queensland section of Gondwana-Land. The Buru–Seram terrane has a basement of low to medium grade metamorphics overlain by siltstones, mudstones and reefal limestones of Triassic age. The reef facies of Seram, the Asinepe Limestone, is treated as allochthonous by Audley-Charles *et al.* (1979). The terrane is believed derived from Papua New Guinea (Pigram & Panggabean 1984). The Obi–Bacan, Bangai–Sula and Buton terranes have basements of metamorphics intruded by late Palaeozoic granitoids on Obi–Bacan and Bangai–Sula, overlain by middle Jurassic shales, and sandstones of Jurassic and Cretaceous age on Obi–Bacan, and Triassic volcanics and shallow marine strata on Bangai–Sula and Buton respectively. The basements of these blocks resemble the eastern part of the Birds Head and Central Papua New Guinea, from where they are believed to be derived

(Pigram & Panggabean 1984). The Sumba continental fragment has a Mesozoic basement of carbonaceous strata with Jurassic and Cretaceous ammonoids and bivalves overlain by Miocene pelagic chalk. It is believed to have its origin on the N.W. Australian margin and was carried north by the spreading of the Argo Abyssal Plain (Chamalaun *et al.* 1982).

SOUTHEAST ASIAN SUTURES AND THEIR AGES

The principal sutures are shown in figure 1. The ages range from Carboniferous to Tertiary reflecting the long and complex accretionary history of the area.

Song Ma and Song Da sutures

The boundary between the Indochina and South China Blocks is marked by the Song Ma–Song Da zone. Various workers recognise two suture zones, the Song Ma and Song Da sutures of Palaeozoic and Mesozoic age respectively (Tran 1979; Gatinsky *et al.* 1985; Sengor 1984, 1986; Sengor *et al.* 1988). Triassic sequences in the Song Da and other Triassic basins of this zone have been regarded as superimposed rift basins by Tran (1979) and Gatinsky *et al.* (1984), but have alternatively been interpreted as an accretionary complex by Sengor (1984). The question as to whether South China and Indochina were welded in the Palaeozoic along the Song Ma suture or in the Triassic along the Song Da zone remains unresolved. Following my previous papers, I here favour the Palaeozoic suturing along the Song Ma suture. The age of the Song Ma suture is indicated to be early Carboniferous by Devonian–early Carboniferous ophiolite and melange ages, early-middle Carboniferous folding and large-scale thrusting and by blanketing strata of middle Carboniferous age. Indochina and South China were therefore probably amalgamated by the middle Carboniferous and together with East Malaya formed 'Cathaysialand' (Gatinsky *et al.* 1984; Lin 1987).

Uttaradit–Nan and Raub–Bentong sutures

These two sutures, regarded as contiguous by most authors, form the boundary between the Sibumasu Block and the Indochina and East Malaya Blocks. The age of these sutures is still contentious. The traditional view is that they are late Triassic in age (Ridd 1980; Mitchell 1981; Sengor 1984) but an earlier suturing in the Permian has been suggested by Helmcke (1985) and in the early Triassic by Mitchell (1989) and Cooper *et al.* (1989). An early Triassic rather than a Permian age is here favoured. The Raub–Bentong suture is also here considered to be probably of early Triassic age. Melange in the Raub area contains limestone clasts of early and late Permian age (Chakraborty & Metcalfe 1987; Metcalfe 1989). Blanketing strata is of late Triassic–Jurassic age and the ages of the S-type collisional Main Range Granites are latest Triassic and early Jurassic. The structural data recently reported from Peninsular Malaysia by Harbury *et al.* (1990) does not necessarily preclude a Triassic suturing of western and eastern Malaya along the Bentong–Raub suture.

Shan boundary suture

Also known as the Mandalay or Sagaing suture it forms the boundary between the Mount Victoria Land Block and Sibumasu. Mitchell (1989) has suggested a late Jurassic or early Cretaceous collision between Mount Victoria Land and Sibumasu. A Cretaceous suturing is here favoured in view of the Cretaceous aged thrusts in the back-arc belt (Mitchell 1990) and the late Cretaceous age for the Western Belt tin-bearing granites. This would also better agree with the age of closure of the Banggong Co-Nu Jiang ocean (Sengor *et al.* 1988).

Woyla suture

Cameron *et al.* (1980) proposed a late Jurassic–early Cretaceous opening of a marginal basin on the southwest margin of Sumatra and the rifting of the Sikuleh and Natal blocks from Sundaland followed by a late Cretaceous closure and suturing. An alternative hypothesis, made by Pulungono (1983) and Barber (1985) involving subduction processes in the Cretaceous and a derivation of the Sikuleh and Natal blocks from Gondwana-Land with a late Cretaceous suturing, is here proposed and is supported by the palaeomagnetic data (see below).

Boyan and Lupar sutures

Recent data from West Kalimantan (Williams *et al.* 1986; Williams & Harahap 1987) indicate a late Cretaceous short-lived subduction which produced the Boyan Melange and which ceased due to clogging by the Semitau Terrane in the late Cretaceous. Subduction then jumped and continued behind the Semitau Block along the Lupar Line producing the Lubok Antu Melange and the Sarawak accretionary prism. The Lubok Antu Melange and accretionary wedge turbidites of the Lupar suture yield early Eocene ages from both clasts and matrix. Tan (1982) and Taylor & Hayes (1983) have suggested that subduction continued into the Miocene but Williams & Harahap have proposed that the Oligocene–early Miocene intrusive rocks in northwest Borneo represent post-subduction intrusions. This would constrain the age of the suture as probably Oligocene. A mid-Miocene age for the suture, consequent upon the arrival of the Luconia terrane, seems more attractive considering that the South China Sea began its spreading in the mid-Oligocene. In that case the I-type small Oligocene and Miocene intrusions of northwest Borneo would have to be related to a north migrated subduction front as suggested by Hamilton (1979). The precise age of the Lupar suture cannot at present be determined but falls in the range late Eocene to mid-Miocene. The age of structural inversion observed in the Malay and West Natuna basins around 25 Ma (Oligocene–Miocene boundary) may be indicative of the cessation of southwest directed subduction beneath Borneo.

Meratus suture

Subduction melange and ophiolite of the Meratus Mountains appears to be of Middle Cretaceous age (Hamilton 1979) and is overlain by Eocene strata. Obduction of ophiolite occurred in the Cenomanian, related to an arc-continent collision (Sikumbang 1986). The Meratus suture is therefore of late Cretaceous age and it may have been a continuation of the Woyla suture in Sumatra.

PALAEOMAGNETIC DATA

All available palaeomagnetic data for Southeast Asia was reviewed by Haile & Briden (1982) and the rather restricted and quality-variable nature of the database was all too evident. Since that time significant contributions to the palaeomagnetic database have been made and an Apparent Polar Wander Path (APWP) has been constructed for the South China Block and palaeolatitudes and senses and amounts of rotation for other terranes are also better known.

Palaeolatitudes

Palaeolatitude plots against time for the South China, Indochina, East Malaya and Sibumasu Blocks are given in figure 2. All four terranes show rapid northward movement during the Permo–Triassic followed by a Jurassic–early Cretaceous southwards drift, a mid-late Cretaceous northwards motion and a Tertiary southwards movement to their present latitudes. The remarkably similar pattern of north–south movements for all four blocks since the Permian suggests that they were in the same plate tectonic regime during that period. It is also clear from these plots that the Sibumasu block travelled rapidly from southern to northern palaeolatitudes during middle Permian–early Triassic times. This is consistent with the proposed early Permian rifting of this block from Australian Gondwana-Land. The sudden southwards movements experienced by all four blocks in the Jurassic–early Cretaceous immediately postdates the collision with North China which took place in the latest Triassic–early Jurassic (Sengor 1985; Lin *et al.* 1985; Lin & Fuller 1990). The database for the other Southeast Asian terranes is too limited to construct palaeolatitude variation diagrams. There are no data for the Mount Victoria Land block. Data from the Semitau terrane include two contrasting Triassic palaeolatitudes of 10.5° S and 17° N (Sunata & Wahyono 1990) and

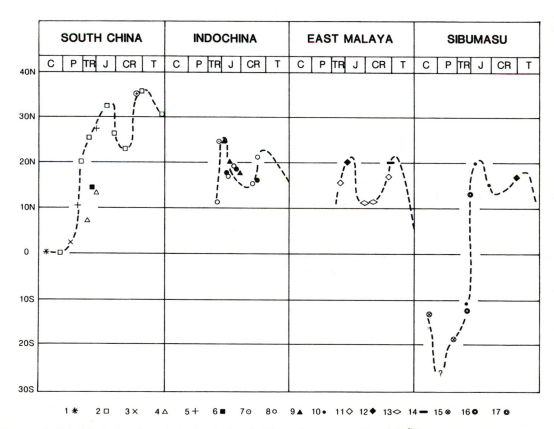

FIGURE 2. Palaeolatitude–time plots for the South China, Indochina, East Malaya and Sibumasu terranes from Carboniferous to Tertiary. Sources of data are: 1, Lin (1987); 2, Lin *et al.* (1985); 3, McElhinny *et al.* (1981); 4, Opdyke *et al.* (1986); 5, Sasajima & Maenaka (1987); 6, Chan *et al.* (1984); 7, Achache *et al.* (1983); 8, Marante & Vella (1986); 9, Achache & Courtillot (1985); 10, Bunopas *et al.* (1989a, b); 11, McElhinny *et al.* (1974); 12, E. A. Schmidtke & M. Fuller (personal communication); 13, Haile & Khoo (1980); 14, Haile *et al.* (1983); 15, Bunopas (1982); 16, Sasajima *et al.* (1978); 17, Haile (1979).

175

equatorial palaeolatitudes from Jurassic to the present (Schmidtke *et al.* 1990). The South West Borneo terrane appears to have been located on or very close to the equator since the Jurassic (Haile *et al.* 1977; Sunata & Wahyono 1990). Palaeomagnetic data for other small terranes in the region is very limited and confined generally to isolated sample sites. Haile (1979) reported some late Triassic and early Cretaceous data from localities in Sumatra which are located on the Sikuleh terrane. These results indicate that the Sikuleh terrane was at 26° S in the late Triassic and at about 10° S in the late Mesozoic. These results are consistent with the proposed derivation from Gondwana-Land in the late Jurassic. Data from West Sulawesi indicates southern palaeolatitudes in the Cretaceous (Sasajima *et al.* 1980; Haile 1978) which support the interpreted derivation from N.W. Australia in the late Jurassic. Data from the Sumba terrane indicate a Jurassic palaeolatitude of 25° S (Otofuji *et al.* 1981) and 14° S in the late Cretaceous (Chamalaun & Sunata 1982). This supports an Australian rather than a Sundaland origin for Sumba. Palaeomagnetic data from the Yaxian terrane of Hainan gives a Cambrian palaeolatitude of 6.7° S which would be consistent with a derivation from Australian Gondwana-Land. An early Permian palaeolatitude of 5 °N for the Echa Formation of the Qiongzhong terrane of Hainan, which contains glacial-marine diamictites, appears to be in conflict with an Australian origin suggested by stratigraphy and palaeobiogeography (Yu 1989).

Rotations

Variations of declination with age indicate sense and amount of rotation for a particular terrane and are potentially very useful in unravelling or constraining palaeotectonics. The Indentor–Extrusion model for Cenozoic tectonics in Southeast Asia (Molnar & Tapponier 1975; Tapponier *et al.* 1982) implies the east and southeast extrusion of the region consequent upon the collision of India with Eurasia and requires significant clockwise rotations for a number of the Southeast Asian terranes. Alternative models (see, for example, Holloway 1981, 1982; Taylor & Hayes 1980, 1983) require counterclockwise rotations for some of the same blocks. It therefore becomes crucial that palaeomagnetic data be assessed to test these models. The South China Block has rotated about 30° clockwise (cw) since the Cretaceous according to Achache *et al.* (1983). The Indochina Block shows evidence of 47° cw rotation since the Lower Cretaceous (Achache *et al.* 1983) and 37° cw rotation since the late Cretaceous (Bunopas *et al.* 1989a). The Sibumasu Block has rotated about 55° cw since the early Cretaceous, much of that rotation occurring around the Cretaceous–Tertiary boundary (Bunopas *et al.* 1989b). This data is in general consistent with the Extrusion model, particularly as these blocks have also moved southwards during the Tertiary (figure 2). However, Cretaceous data from the South West Borneo Block, the Semitau Block and from East Malaya all show counterclockwise (ccw) rotations of between 30° and 50° since the late Cretaceous (Haile 1979; Schmidtke *et al.* 1990; Sunata & Wahyono 1990). ccw rotation of Luzon in the Eocene–Oligocene followed by rapid ccw rotation in the Miocene are reported by Fuller *et al.* (1983) and Fuller *et al.* (1990) which appears to support the Holloway model. The ccw rotations for Borneo and East Malaya also appear to contradict the Indentor Extrusion model. Results of ongoing palaeomagnetic studies in the region by Fuller and co-workers should provide more constraints on the Cenozoic tectonic models for Southeast Asia and these are eagerly awaited.

ALLOCHTHONOUS TERRANE PROCESSES IN SE ASIA

Apparent Polar Wander Paths

Comparisons of APWPs of the Southeast Asian Blocks with each other and with APWPs of Eurasia and Gondwana-Land should allow the timings of rifting and amalgamation or accretion to be constrained. Unfortunately, an insufficient database for Southeast Asia does not permit such comparisons at present. Only the South China Block has a fairly well-constrained APWP (Lin *et al.* (1985) and comparisons are hence limited to individual poles from Southeast Asia with APWPs for Eurasia, Gondwana-Land and South China.

ALLOCHTHONOUS TERRANE PROCESSES

Fundamental factors in elucidating the terrane processes of the region and for constraining palaeogeographic maps are the timings of rifting of terranes from their parent cratons, the positions and relative rotations of terranes during their pre-rift, drift, and post-drift histories, the timings of suturing (amalgamation and accretion) and post-suturing structural and tectonic modifications. Criteria used to identify or constrain the timings and positions of terranes are varied in type and reliability. The main criteria used in timing the rifting of terranes are ocean floor ages and magnetic stripe data, divergence of APWPs and palaeolatitudes from continental palaeomagnetism, age of associated rift volcanics and intrusions, regional unconformities, ages of major block faulting episodes and associated slumping, development of different biotic provinces on separating blocks. The palaeopositions of continental terranes during their drift may be estimated from well-constrained palaeomagnetism which will give palaeolatitudes and sense and amount of rotation relative to present. Stratigraphical and palaeontological data may also be indicative of palaeolatitude or proximity to other continental regions. Many criteria have been used to date the suturing of one terrane to another including ophiolite obduction ages, ages of melanges, ages of stitching plutons, ages of subduction-related and collisional-related plutons, ages of and chemical or isotopic changes in volcanic arcs, convergence of APWPs, loops or disruptions in individual APWPs, age of blanketing strata, palaeobiogeography, structural geology (e.g. age of thrusting, nappe formation), stratigraphic sequences (e.g. pelagic-flysch-molasse). Application of the above criteria allows the rift–drift–suture processes for the Southeast Asian terranes to be worked out and these are discussed below.

Palaeozoic processes

The first event in the history of the Southeast Asian terranes was the Palaeozoic rifting of South China, Indochina, East Malaya and S.W. Borneo from Gondwana-Land. The precise age of rifting is not known but thought to be Silurian or Devonian by Lin (1987, this Symposium). The presence of a regional unconformity of late Devonian–early Carboniferous age supports a Devonian rifting. These blocks then amalgamated by late early Carboniferous times to form a superterrane referred to as the East Asian Continent by Gatinsky *et al.* (1984) and Cathaysialand by Lin (1989) on which the Cathaysian flora developed during the Carbo–Permian. This superterrane was located in low northern equatorial latitudes during the Carboniferous (Lin 1987; see figure 3*a*). In the late early to Middle Permian, a major rifting phase occurred on the northeast Gondwana-Land margin (Powell 1976; Bird 1987; Audley-Charles 1988) indicating that a substantial continental fragment (or fragments) separated from Gondwana-Land at that time. Audley-Charles (1988) suggested that the blocks rifting at this

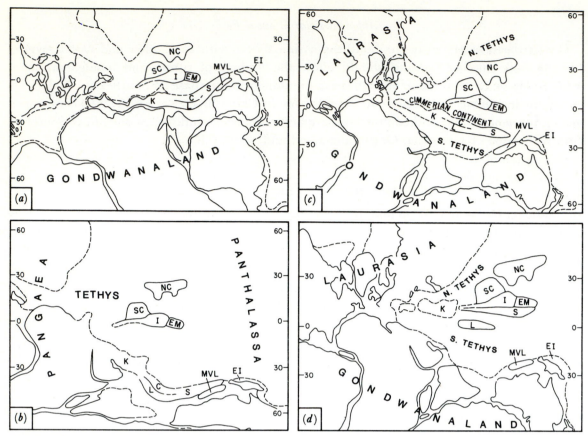

FIGURE 3. Palaeogeographic reconstructions for (*a*) early Carboniferous, (*b*) early early Permian, (*c*) middle–late Permian and (*d*) late Triassic. Based on the reconstructions of Smith *et al.* (1981) and partly after Metcalfe (1988). Present day outlines are for reference only. NC = North China, SC = South China, I = Indochina, EM = East Malaya, K = Kreios, C = Changtang, L = Lhasa, S = Sibumasu, MVL = Mount Victoria Land, EI = Eastern Indonesian terranes.

time were Iran, North Tibet (Changtang) and Indochina, thus leaving Sibumasu as a source for the Triassic intracratonic basin deposits seen in Timor. Metcalfe (1988), however, suggested that it was in fact the Sibumasu, Lhasa and Changtang blocks (along with other terranes that comprised the Cimmerian continent of Sengor) which rifted from Gondwana at that time. That view is still held and the source for the Triassic intracratonic basin sediments in Timor is believed to be Mount Victoria Land. Sibumasu must have remained attached or close to N.W. Australia until the early Permian (figure 3*b*) as indicated by early Permian glacial-marine deposits, cool-water faunas and faunas with N.W. Australian affinities (Metcalfe 1988). The middle–late Permian and early Triassic faunas of Sibumasu show affinities to Cathaysialand and to northern Tethys province types (Metcalfe 1988*a, b*) indicating that Sibumasu had already rifted from Gondwana-Land. This is supported by the palaeomagnetic data. During its travel northwards across the Tethys, the Cimmerian continent fragmented both longitudinally and latitudinally separating the Lhasa and Changtang–Sibumasu blocks and separating the Western Cimmerian Continent in Turkey and Iran from the Southeast Asian and Tibetan terranes (figure 3*c*). In late Permian times, Sibumasu began its collision with Cathaysialand and the suturing to Indochina and East Malaya was largely completed by the early Triassic (see above).

ALLOCHTHONOUS TERRANE PROCESSES IN SE ASIA

Mesozoic processes

Granites generated during or post the collision between Sibumasu and Indochina–East Malaya were emplaced in the latest Triassic to early Jurassic (Liew & Page 1985; Cobbing *et al.* 1986; Darbyshire 1987) which would indicate an early Triassic or earlier age for the collision considering that there may be up to 30 Ma time difference between the generation and emplacement of collisional granites (England & Thompson 1986). The late Triassic–early Jurassic period saw the collision of Cathaysialand with North China and an overall consolidation of the Southeast Asian cratonic 'core'. The Indosinian orogeny was the result of this collision and final consolidation. The Triassic basins were largely closed at this time and continental deposits were widespread over large parts of the Southeast Asian craton during the Jurassic and Cretaceous. The final suturing of mainland Southeast Asia to Eurasia probably occurred in the early–middle Jurassic. A further period of rifting occurred on the north-eastern margin of Gondwana-Land during the late Jurassic and another sliver of the cratonic margin separated at this time. This sliver included the Mount Victoria Land, Sikuleh, Natal, Mangkalihat, West Sulawesi and Sumba terranes along with other unidentified continental crust (including the Banda allochthon of Audley-Charles?). As Neotethys (Tethys III) opened up in the late Jurassic, the Banggong Co-Nu Jian ocean (Tethys II) continued to subduct beneath Eurasia and the Lhasa block finally collided with the Changtang and Sibumasu terranes in the late Jurassic–early Cretaceous (figure 4a). The Mount Victoria Land, Sikuleh and Natal terranes continued their northward travel during the Cretaceous (figure 4b) and had sutured to Sibumasu by the late Cretaceous (figure 4c). A short-lived subduction in the late Cretaceous brought together the Semitau and South West Borneo terranes along the Boyan Suture and the Philippine arc was in its incipiancy. There was also a significant clockwise rotation of 'Sundaland' during the late Cretaceous and India travelled rapidly northwards towards its collision with Eurasia.

Cenozoic processes

During the last decade, the Indentor–Extrusion model for the Cenozoic of Southeast Asia (Molnar & Tapponier 1975; Tapponier *et al.* 1982; Tapponier *et al.* 1986) achieved wide acceptance amongst geologists working in the region. The model proposes the migration of a prograding zone of deformation across Asia which occurs concurrently with the northward moving India–Eurasia collision front and the successive activation of several large left-lateral strike–slip faults. During the late Eocene to Miocene (40–20 Ma), 'Sundaland' was pushed sideways and extruded to the east and southeast rotating clockwise by about 25° during the Oligocene to early Miocene simultaneous with the opening of the South China Sea. Later in the Miocene, Tibet and China moved several hundred kilometres to the east and the opening of the South China Sea stopped but 'Sundaland' continued to rotate clockwise a further 40° in sympathy with the South China Block. The opening of the Andaman Sea was proposed to be the result of the eastwards motion of the Southeast Asian Blocks relative to India and as a possible counterpart to the South China Sea (Tapponier *et al.* 1986). The main large left-lateral strike–slip faults that have accommodated these movements in Southeast Asia have been identified as the Altyn Tagh-Kun Lun and Red River faults. The main test of the extrusion model comes from recognition of major left-lateral movements along these faults and in

179

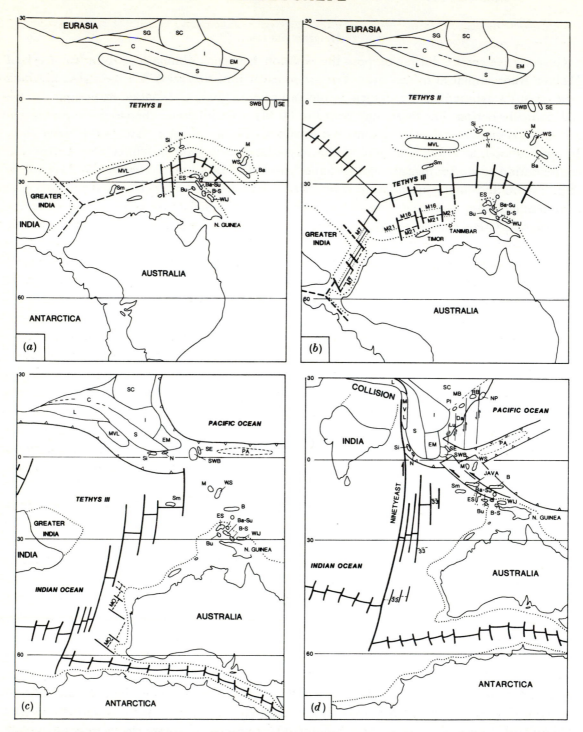

FIGURE 4. Palaeogeographic reconstructions for (a) late Jurassic, (b) early Cretaceous, (c) late Cretaceous and (d) late Eocene. SG = Songpan Gangzi accretionary complex, SWB = South West Borneo, SE = Semitau, Si = Sikuleh, N = Natal, M = Mangkalihat, WS = West Sulawesi, Ba = Banda Allochthon, ES = East Sulawesi, O = Obi-Bacan, Ba-Su = Bangai-Sula, Bu = Buton, B-S = Buru-Seram, WIJ = West Irian Jaya. Other terrane symbols as in figure 3. Partly after Smith *et al.* (1981), Audley-Charles (1988) and Audley-Charles *et al.* (1989). Present day outlines are for reference only.

demonstrating the necessary progressive clockwise rotations and southeast movements of the Southeast Asian Blocks.

Alternative models have been put forward for the South China Sea region by Taylor & Hayes (1980, 1983) and Holloway (1981, 1982) where the South China Sea develops due to rifting of the South China–Indochina margin and consumption of Mesozoic oceanic crust by southwards subduction beneath Borneo and the Pacific plate. They also propose that substantial continental crust attenuation took place on the South China margin between the latest Cretaceous and earliest Palaeocene. This rift onset would be too old to be due to the collision of India with Eurasia. They also propose that the spreading in the South China Sea was essentially north–south and that the Philippine arc and Borneo underwent significant counterclockwise rotations. The North Palawan, Reed Bank and Luconia–Dangerous Ground–Spratley Islands terranes travelled south during Mid-Oligocene to early Miocene times. Cessation of subduction to the south in Mid-Miocene times is indicated by an unconformity on Palawan and Holloway (1982) has proposed diachronieity of the suture as collision of blocks with Borneo progressively inactivates the subduction zone from west to east. Hence, the Taylor & Hayes–Holloway model proposes an essentially local closed system for the South China Sea.

Tapponier et al. (1986), however, consider the South China Basin an Atlantic-type marginal basin bounded by passive continental margins to the north and south. Clockwise rotations of the South China (30°), Indochina (47°), and Sibumasu (55°) since the Cretaceous (see above) are consistent with the extrusion model. Borneo, the Philippines and East Malaya however show clear counterclockwise rotations since the Cretaceous (see above) which support the Taylor & Hayes–Holloway model. Further detailed palaeomagnetic work is required, along with structural data on strike–slip faults to constrain the models for the Cenozoic evolution of Southeast Asia. What is clear, however, is that the North Palawan, Reed Bank and Luconia–Dangerous Ground–Spratley terranes were detached from the South China margin in the Middle Oligocene, travelled south during the spreading of the South China Sea and collided with, and/or were underthrust beneath, Borneo and Palawan during the Miocene (Hinz & Schluter 1983; Mohammad et al. 1987; Tan & Lamy 1989).

Fragments of the North Australian margin (Mangkalihat, West Sulawesi, etc.) that rifted in the Jurassic continued to move northwards during the Cenozoic along with Australia and were translated north and west by major transcurrent faults in a kind of bacon-slicer mechanism between the Australian and Pacific plates (figure 4d). This mechanism also moved fragments of the New Guinea margin of Australia (West Irian Jaya, Buru-Seram, Buton, Bangai-Sula, East Sulawesi) westwards along major fault dislocations (the Sorong fault being a recently active one). Proto-Borneo was formed when the Mangkalihat and Paternoster terrane (then part of the West Sulawesi block) sutured to the other Borneo blocks along the Adio and Meratus sutures. The collision between Southeast Asia and the Australian craton began with the collision of West and East Sulawesi at around 15 Ma (late Miocene) and continued with the collisions of Seram and Timor at 5 and 3 Ma respectively (Audley-Charles et al. 1988).

REFERENCES

Achache, J. & Courtillot, V. 1985 Earth planet. Sci. Lett. **73**, 147–157.
Achache, J., Courtillot, V. & Besse, J. 1983 Earth planet. Sci. Lett. **63**, 123–136.
Audley-Charles, M. G. 1983 Nature, Lond. **306**, 48–50.

Audley-Charles, M. G. 1988 In *Gondwana and Tethys* (ed. M. G. Audley-Charles & A. Hallam), pp. 79–100. Geological Society Special Publication no. 37.

Audley-Charles, M. G., Ballantyne, P. D. & Hall, R. 1988 *Tectonophysics* **155**, 317–330.

Audley-Charles, M. G., Barber, A. J., Norvick, M. S. & Tjokrosapoetros, S. 1979 *J. geol. Soc. Lond.* **136**, 547–568.

Barber, A. J. 1985 In *Tectonostratigraphic terranes of the Circum-Pacific region* (ed. D. G. Howell), pp. 523–528. Circum Pacific Council for Energy and Mineral Resources, Earth Science Series, no. 1.

Bird, P. 1987 *The geology of the Permo–Triassic rocks of Kekneno, west Timor.* Ph.D. thesis, University of London, U.K.

Bunopas, S. 1982 *Geol. Surv. Pap. no. 5*, Department of Mineral Resources, Thailand. (810 pages.)

Bunopas, S., Marante, S. & Vella, P. 1989 *a* In *4th Int. Symp. pre-Jurassic evolution of East Asia, IGCP Project 224. Rep. Abstr.* **1**, 63–64.

Bunopas, S., Marante, S. & Vella, P. 1989 *b* In *4th Int. Symp. pre-Jurassic evolution of East Asia, IGCP Project 224. Rep. Abstr.* **1**, 61–62.

Burrett, C. & Long, J. 1989 In *4th Int. Symp. pre-Jurassic evolution of East Asia, IGCP Project 224. Rep. Abstr.* **1**, 83.

Burrett, C. & Stait, B. 1986 *Geol. Soc. Malaysia Bull.* **19**, 103–107.

Cameron, N. R., Clarke, N. C. G., Aldiss, D. T., Aspden, T. A. & Djunuddin, A. 1980 *The geological evolution of northern Sumatra*, Ninth Indonesian Petroleum Convention, Jakarta. (53 pages.)

Chakraborty, K. R. & Metcalfe, I. 1987 *Warta Geologi* **13**, 62–63.

Chamalaun, F. H. & Sunata, W. 1982 *CCOP Tech. Pap.* **13**, 162–194

Chamalaun, F. H., Grady, A. E., von der Borch, C. C. & Hartono, H. M. S. 1982 *Am. Ass. Petrol. Geol. Mem.* **34**, 361–376.

Chan, L. S., Wang, C. Y. & Wu, X. Y. 1984 *Geophys. Res. Lett.* **11**, 1157–1160.

Charvet, J. & Faure, M. 1989 In *4th Int. Symp. pre-Jurassic evolution of East Asia, IGCP Project 224. Rep. Abstr.* **1**, 19–20.

Cobbing, E. J., Mallick, D. I. J., Pitfield, P. E. J. & Teoh, L. H. 1986 *J. Geol. Soc. Lond.* **143**, 537–550.

Cooper, M. A., Herbert, R. & Hill, G. S. 1989 In *Proc. Int. Symp. Intermontane Basins: Geology and Resources* (ed. T. Thanasuthipitak & P. Ounchanum), pp. 231–242. Chiang Mai University, Thailand.

Darbyshire, D. P. F. 1987 *Warta Geologi* **13**, 117–120.

England, P. C. & Thompson, A. 1986 In *Collision tectonics* (ed. M. P. Coward & A. C. Ries), pp. 83–94. Geological Society Special Publication no. 19.

Fuller, M., Haston, R. & Almasco, J. 1990 *Tectonophysics.* (In the press.)

Fuller, M., McCabe, R., Williams, I. S., Almasco, J., Encina, R. Y., Zanoria, A. S. & Wolfe, J. A. 1983 In *The tectonic and geologic evolution of Southeast Asian seas and islands* (ed. D. E. Hayes), pp. 79–94. A.G.U., Geophys. Monograph no. 27.

Gatinsky, Y. G., Hutchison, C. S., Minh, N. N. & Tri, T. V. 1984 *27th Int. Geol. Congress, Moscow, Rep.* **5**, 225–239.

Haile, N. S. 1978 *Tectonophysics* **46**, 77–85.

Haile, N. S. 1979 *a* *J. Geol. Soc. Lond.* **136**, 541–545.

Haile, N. S. 1979 *b* *Warta Geologi* **5**, 19–22.

Haile, N. S. & Briden, J. C. 1982 In *Palaeomagnetic research in Southeast and East Asia* (ed. J. C. Briden), *CCOP Tech. Pap.* **13**, 25–46.

Haile, N. S. & Khoo, H. P. 1980 *Geol. Soc. Malaysia Bull.* **12**, 75–78.

Haile, N. S., McElhinny, M. W. & McDougall, I. 1977 *J. geol. Soc. Lond.* **133**, 133–144.

Haile, N. S., Beckinsale, R. D., Chakraborty, K. R., Abdul, H. H. & Tjahjo, H. 1983 *Geol. Soc. Malaysia Bull.* **16**, 71–85.

Hamilton, W. 1979 *US Geol. Surv. Prof. Pap.* **1078**, 1–345.

Harbury, N. A., Jones, M. E., Audley-Charles, M. G., Metcalfe, I. & Mohamad, K. R. 1990 *J. geol. Soc. Lond.* **147**, 11–26.

Helmcke, D. 1983 *Earth Evol. Sci.* **4**, 309–319.

Helmcke, D. 1985 *Geol. Rdsch.* **74**, 215–228.

Hinz, K. & Schluter, P. 1983 *Bundersanstalt fur Geowissenschaften und Rohstoffe* 95701.

Holloway, N. H. 1981 *Geol. Soc. Malaysia Bull.* **14**, 19–58.

Holloway, N. H. 1982 *Bull. Am. Ass. Petrol. Geol.* **66**, 1355–1383.

Hsu, K. J., Sun, S., Li, J., Chen, H., Pen, H. & Sengor, A. M. C. 1988 *Geology* **16**, 418–421.

Hutchison, C. S. 1989 *Geol. Soc. Malaysia Bull.* **20**, 201–220.

Klimetz, M. P. 1987 In *Terrane accretion and orogenic belts* (ed. E. C. Leitch & E. Scheibner), pp. 221–134. A.G.U. Geodynamic Series, vol. 19.

Liew, T. C. & McCulloch, M. T. 1985 *Geochimica Cosmochimica Acta* **49**, 587–600.

Liew, T. C. & Page, R. W. 1985 *J. Geol. Soc. Lond.* **142**, 515–526.

Lin, J. 1987 *11th International Congress of Carboniferous Stratigraphy and Geology, Abstracts* **1**, 283.

Lin, J. & Fuller, M. 1990 *Nature, Lond.* (In the press.)

Lin, J., Fuller, M. & Zhang, W. 1985 *Nature, Lond.* **313**, 444–449.

Marante, S. & Vella, P. 1986 *J. southeast Asian Earth Sci.* **1**, 23–31.

McElhinny, M. W., Haile, N. S. & Crawford, A. R. 1974 *Nature, Lond.* **252**, 641–645.

McElhinny, M. W., Embleton, B. J. J., Ma, X. H. & Zhang, Z. K. 1981 *Nature, Lond.* **293**, 212–216.

Metcalfe, I. 1985 *Mem. Soc. Geol. France* **147**, 107–118.

Metcalfe, I. 1986 *Geol. Soc. Malaysia Bull.* **19**, 153–164.

Metcalfe, I. 1988*a* In *Gondwana and Tethys* (ed. M. G. Audley-Charles & A. Hallam), pp. 101–118. Geol. Soc. London Spec. Pub. no. 37.

Metcalfe, I. 1988*b* *Warta Geologi* **14**, 34.

Metcalfe, I. 1989 In *Proc. Int. Symp. Intermontane Basins: Geology and Resources* (ed. T. Thanasuthipitak & P. Ounchanum), pp. 173–186, Chiang Mai University, Thailand.

Mitchell, A. H. G. 1981 *J. geol. Soc. Lond.* **138**, 109–122.

Mitchell, A. H. G. 1989 In *Tectonic evolution of the Tethyan region* (ed. A. M. C. Sengor), pp. 567–583. Dordrecht: Kluwer.

Mitchell, A. H. G. 1990 *Geol. Soc. Malaysia Bull.* **20**.

Mohammad, A. M., Ismail, M. I. & Hinz, K. 1987 *Geol. Soc. Malaysia Petroleum Geol. Seminar, Abstr.* **19**.

Molnar, P. & Tapponier, P. 1975 *Science, Wash.* **189**, 419–426.

Opdyke, N. D., Huang, K., Xu, G., Zhang, W. Y. & Kent, D. V. 1986 *J. geophys. Res.* **91**, 9553–9568.

Otofuji, Y., Sasajima, S., Nishimura, S., Dharma, A. & Hehuwat, F. 1981 *Earth planet. Sci. Lett.* **52**, 93–100.

Pigram, C. J. & Panggabean, H. 1984 *Tectonophysics* **107**, 331–353.

Powell, D. E. 1976 *J. Austral. Petrol. Explor. Assoc.* **16**, 13–23.

Pulunggono, A. 1983 *Sistem Sesar Utama dan Pembentukan Cekungan Palembang*. Ph.D. thesis, Institute of Technology, Bandung. (239 pages.)

Pulunggono, A. & Cameron, N. R. 1984 *Proc. 13th A. Convention Indonesian Petroleum Assoc.* **1**, 121–143.

Ren, J. 1989 *4th Int. Symp. pre-Jurassic evolution of East Asia, IGCP Project 224. Rep. Abstr.* **1**, 18.

Ridd, M. F. 1980 *J. geol. Soc. Lond.* **137**, 635–640.

Rutten, M. G. 1940 *Proc. Nederl. Akademie Van Wetenschappen* **18**, 1061–1064.

Sasajima, S. & Maenaka, K. 1987 *IGCP Project 224, Pre-Jurassic geological evolution of eastern continental margin of Asia, Rep. no. 2*, 139–150.

Sasajima, S., Nishimura, S., Hirooka, K., Otofuji, Y., Vanleeuwan, T. & Hehuwat, F. 1980 *Tectonophysics* **64**, 163–172.

Sasajima, S., Otofuji, Y., Hirooka, K., Suparka, Suwijanto & Hehuwat, F. 1978 *Rock Magnetism Paleogeophysics* **5**, 104–110.

Schmidtke, E. A., Fuller, M. & Haston, R. B. 1990 *Tectonics*. (In the press.)

Sengor, A. M. C. 1984 *Geol. Soc. Am. Spec. Pap.* **195**. (82 pages.)

Sengor, A. M. C. 1985 *Nature, Lond.* **318**, 16–17.

Sengor, A. M. C. 1986 *Tectonophysics* **127**, 177–195.

Sengor, A. M. C. 1989 In *Tectonic evolution of the Tethyan region* (ed. A. M. C. Sengor), pp. 1–22. Dordrecht: Kluwer.

Sengor, A. M. C. & Hsu, K. J. 1984 *Mem. Soc. Geol. France* **147**, 139–167.

Sengor, A. M. C., Altiner, D., Cin, A., Ustaomer, T. & Hsu, K. J. 1988 In *Gondwana and Tethys* (ed. M. G. Audley-Charles & A. Hallam), pp. 119–181. Geol. Soc. Spec. Publ. no. 37.

Sikumbang 1986 Ph.D. thesis, University of London, U.K.

Smith, A. G., Hurley, A. M. & Briden, J. C. 1981 *Phanerozoic palaeocontinental world maps*. Cambridge University Press. (102 pages.)

Sunata, W. & Wahyono, H. 1990 International Oceanographic Committee, SEATAR Transects, Transect VII.

Tan, D. N. K. 1982 *Geol. Soc. Malaysia Bull.* **15**, 31–46.

Tan, D. N. K. & Lamy, J. M. 1989 *Geol. Soc. Malaysia Petrol. Geology Seminar Abstr.* 22–21.

Tapponier, P., Peltzer, G. & Armijo, R. 1986 In *Collision tectonics* (ed. M. P. Coward & A. C. Ries), pp. 115–157. Geol. Soc. Spec. Publ. no. 19.

Tapponier, P., Peltzer, G., LeDain, A. Y., Armijo, R. & Cobbold, P. 1982 *Geology* **10**, 611–616.

Taylor, B. & Hayes, D. E. 1980 In *The tectonic and geologic evolution of Southeast Asian seas and islands* (ed. D. E. Hayes), pp. 23–56. Am. Geophys. Union Geophys. Monograph no. 23.

Taylor, B. & Hayes, D. E. 1983 In *The tectonic and geologic evolution of Southeast Asian seas and islands, Part 2* (ed. D. E. Hayes), pp. 23–56. Am. Geophys. Union Geophys. Monograph no. 27.

Tran, V. T. 1979 *Geology of Vietnam (North Part)*. General Geological Department, Hanoi, Vietnam.

Wang, H. 1986 In *The geology of China* (ed. T. Yang), pp. 235–276. Oxford: Clarendon Press.

Williams, P. R. & Harahap, B. H. 1987 *Austral. J. Earth Sci.* **34**, 405–415.

Williams, P. R., Supriatna, S. & Harahap, B. 1986 *Geol. Soc. Malaysia Bull.* **19**, 69–78.

Yu, Zi-ye 1989 *4th Int. Symp. pre-Jurassic evolution of East Asia, IGCP Project 224. Rep. Abstr.* **1**, 11–12.

Discussion

E. Irving, F.R.S. (*Pacific Geoscience Centre, Sydney, Canada*). How does Dr Metcalfe distinguish between north and south latitudes determined palaeomagnetically from Triassic rocks?

I. METCALFE

I. METCALFE. There are some Jurassic and Cretaceous results that suggest that the rotation of Peninsular Malaysia reverses between the Triassic and Cretaceous. These results are from dykes in eastern Malaya and from sediments in the Tekai region of Pahang. If this interpretation is correct, then the polarity is determined and the Malay Peninsular is in the Northern Hemisphere. This problem is not finally resolved and is the subject of present studies.

General discussion

A. TRENCH (*Department of Earth Sciences, University of Oxford, U.K.*). Several participants at the meeting commented upon the impact of the terrane hypothesis in directing geological thinking to consider strike–slip movements within the British Caledonides. It is perhaps pertinent to recount its effect on geophysical research.

The terrane hypothesis essentially makes two geophysically testable predictions. These are (i) that crustal fragments of disparate geographical origin are juxtaposed in the final crustal collage, and (ii) that terrane movements bring together blocks of different crustal character.

The former of these predictions lends itself to palaeomagnetic study whereas the latter might be addressed using a combination of seismic, electrical and potential field methods.

With regard to palaeomagnetism, analogies drawn between the Caledonides and the western North American Cordillera focused attention upon the respective rotation histories of individual Caledonian terranes. Several local rotations have since been identified (see, for example, Smethurst & Briden 1988; Trench *et al.* 1988) although whether these affect entire terranes is as yet unclear. If these rotations can be considered as evidence favouring the presence of strike–slip movements, other evidence suggests strike–slip movements to be of limited importance. Most notably, a combined Palaeozoic Apparent Polar Wander Path (APWP) from the northern British terranes is found to proxy for that of the North American craton after closing the present Atlantic. Their similar rotation would then argue against major strike–slip movement (which is no longer required by the palaeolatitudinal data (see Stearns *et al.* 1989). Obviously these observations are not mutually exclusive and can be accommodated by an orogen in which strike–slip played an influential, but not universal role. A further understanding awaits the identification of additional rotated terranes and a better time calibration of the compared APWPs.

REFERENCES

Smethurst, M. A. & Briden, J. C. 1988 Palaeomagnetism of Silurian sediments in W Ireland: evidence for block rotation in the caledonides. *Geophys. J.* **95**, 327–346.

Stearns, C., Van der Voo, R. & Abrahamsen, N. 1989 A new Siluro–Devonian palaeopole from early Palaeozoic rocks of the Franklinian Basin, North Greenland fold belt. *J. geophys. Res.* **94**, 10669–10683.

Trench, A., Bluck, B. J. & Watts, D. R. 1988 Palaeomagnetic studies within the Ballantrae Ophiolite; southwest Scotland: magnetotectonic and regional implications. *Earth planet. Sci. Lett.* **90**, 431–448.

Concluding remarks

By J. F. Dewey, F.R.S.

Department of Earth Sciences, Parks Road, Oxford OX1 3PR, U.K.

During the course of this Discussion Meeting, a very large amount of regional tectonic geology was displayed, and debated critically in a terrane framework, on scales ranging from the whole of the North American Precambrian or the Mesozoic–Cenozoic Tethys down to particular segments of the Caledonides and Alpides. A wide spectrum of opinion was expressed from those who believe that the terrane methodology is a critical and essential objective stage in data handling before any rational palaeogeographic and palaeotectonic synthesis can be attempted in plate boundary zones to those who believe that the terrane philosophy is fundamentally flawed, dangerous, and pernicious, in that it leads to random data collection and the obscuring of fundamental plate tectonic processes. Another view was that terranology has been useful in drawing our attention to the importance of large pre-collisional strike–slip or transform motions in orogenic belts and the juxtaposition of disparate elements and zones. Yet another position was that there is nothing new in terranology that is not implicitly and explicitly inherent in plate boundary processes and that terrane analysis is simply another harmless word for what most careful regional geological synthesizers have been doing since the early 1970s. Naturally, no coherent consensus view emerged from the discussion, but an important result was that a huge amount of excellent regional and global geology and tectonic ideas were discussed in the context of the problems and complexities of plate boundary zone evolution and the mechanisms by which objects from the size of 'knockers' to continents, detach, move and weld to form collages at all scales.

Davy Jones, one of the originators of terranology in the North American Cordillera, elegantly expounded the terrane methodology with many detailed examples and emphasized that terrane analysis is not an end in itself but merely a 'holding operation' for a particular terrane until its provenance can be established. Jones argued that the 'home port' of any major Cordilleran terrane has not been identified unequivocally, even those that have a North American cratonic margin provenance. Faunal studies suggest that some terranes, previously thought to have originated near the Cordilleran margin were situated thousands of kilometres to the west and were separated from the continent by broad ocean basins. Jones elaborated his view that only fragments of plate tectonic environments can be recognized in the Cordilleran terrane collage and that all attempts to draw cohesive plate tectonic cross sections across the Cordillera, even on a limited scale is impossible. This position was supported broadly, in the northern Cordillera by Irving and Wynne from palaeomagnetic data, although they argued that the various elements of the western Cordillera in Canada were always close to North America, moving first south, then north. This view of the Cordillera was challenged by Gary Ernst and Warren Hamilton, who, while admitting the clear evidence of great mobility, believe that sensible palaeotectonic coherence exists between some Cordilleran elements on a fairly large scale. They cited, particularly, the Franciscan Assemblage, the Great Valley and the Sierra Nevada as an originally laterally and genetically related subduction-accretion complex, forearc basin and calc-alkaline arc, a view supported in discussion by Alastair Robertson. Ernst argued against a haphazard accretion of Cordilleran terranes because a gradual sectorial enlargement towards the modern continental margin is indicated by coherent broadly continuous isotopic–geochemical provinces from scales of entire physiographic province to that of quadrangle. Ernst suggested further that, metamorphic and structural discontinuities reflect movements among closely associated, co-evolving lithotectonic units and do not imply the juxtaposition of exotic genetically unrelated terranes. Hamilton supported this view and also took the more aggressive stance that few of those who attempt terrane analysis have sufficient expertise in and knowledge of the characteristics of modern convergent margins, of the petrology of modern magmatic provinces, of global patterns of sedimentation and palaeobiogeography with the result that the literature is cluttered with simplistic assumptions and ill-constrained speculations.

Plate boundary evolution, within which framework terranes must be understood, is immensely complicated for several reasons. First, plate kinematics is inherently complicated; slip vectors must continuously change and triple junctions evolve and migrate along plate boundaries. Thus plate boundaries cannot be simple single narrow fault zones across which motions remain constant. Secondly, the vertical rheological structure of plate boundary zones is a response to thermo-mechanical structure. Brittle layers that fragment rotate and displace may alternate with ductile layers in which continuum mechanics is more important. The rock response to plate displacement is also dominated by lithospheric zones in which the vertically integrated strength is lowest such as the axes of volcanic arcs and the zones just inboard of rifted continental margins analysed by Steckler. Thirdly, plate slip vectors in the continental lithosphere are, generally, converted and partitioned into strain and displacement in extremely complicated ways, such that principal displacement zones change with time. Fourthly, direct and simple relationships between plate slip vector and plate boundary zone structure, may be modified and obscured by local body force mechanisms. Orogenic collapse and subduction roll-back are driven by topographic gradients, variations in crustal and lithospheric thickness, and the sink rate of subducting slabs and may generate local slip directions and rates that are at azimuthal variance and much faster than plate relative slip vectors. This may allow very fast, but localized slip and strain rates that allow very rapid terrane motion for short distances and short periods.

A particular difficulty lies in a semantic problem of terranology. If we define a terrane as a geographic unit with a distinctive tectono-stratigraphic assembly, what are the minimum and maximum size limits? Discarding the motion of boulders and sand grains, is there a scale at which terrane becomes a useful word from the upper extreme of a continent like peninsular India to the lower limit of a knocker in a mélange via units such as microplates, microcontinents, blocks slivers and flakes? I believe that, although rigorous size definitions are not useful, the term terrane does not become useful above the scale of a few hundred kilometres or below a few kilometres. Its particularly useful function is in denoting palaeogeographic uncertainty; continents, even pre-Mesozoic ones, can be palaeogeographically manipulated successfully and the prevalence of knockers in mélanges and shear zones is bound to remain an obviously difficult problem. Least of all, the term terrane is not useful in denoting degree of allochthoneity, as a supposed novel methodology, or as a process that somehow replaces and goes beyond plate tectonics. Terranology, if it is useful at all, is only relevant as an obvious consequential subset of plate boundary kinematics as a concept involving the rearrangement of crustal and lithospheric fragments. There is a particular danger in the confused usage of the word terrane, which of course is scarcely the fault of those who would use the term in a 'rigorous' way. During the meeting, the dangers of the term emerged in such phrases and statements as 'Himalayan Terrane', 'an unknown terrane to the west', and 'we really don't need to use the word terrane'. In this context, the real value, or perhaps lack of value, of the terrane concept was impressed upon me by listening to Davy Jones's excellent exposition of the way in which the Cordilleran belt might have evolved. Jones spoke, not of the mindless definition of terranes, but about the genetic significance of tectono-stratigraphic assemblages and how they might have been related; his lecture was almost a denial of so-called 'objective' terranology and a lesson to the sycophantic followers of the terrane concept.

Terranology occupies a wholly different role and significance in different orogenic belts depending on the size and history of the oceanic complex whose subduction and/or demise led to the belt's development and on the plate slip rates and duration along its margins. The North American Cordillera has lain along the margin of a plate tectonic 'shredding machine' throughout at least the Mesozoic and Cenozoic and probably throughout the Palaeozoic since the late Precambrian extensional event that generated the western North American rifted margin. During the late Mesozoic and Cenozoic, the shredding has been principally dextral just as the shredding was mainly sinistral along the northwest Iapetus margin of the Caledonides from middle Ordovician to middle Devonian times. A large ocean, and lots of space and time is likely to lead to substantial terrane motions along its margin. In contrast, short-lived smaller oceans are less likely to generate large terrane motions caught up in the final collisional collage. Thus, in the western region of the European Tethysides, terrane motion was severely constrained in a concertina mode as argued by Sengör and Yilmaz in Turkey. To the east, as the Tethyan oceanic complex widened, large terrane motions are indicated by palaeomagnetic (Lin and Fuller) and geological (Metcalfe, Audley Charles and Harris) data.

If terrane mapping is followed as a mindless procedure that portrays an orogenic zone as an undecipherable, model-independent collage of disparate elements, little progress can be made. If, on the other hand, we look for 'non-terraned' segments of the orogen that may be used as a template to decipher terrane assemblages with excision and repetition elsewhere along strike, substantial progress can be made. The Appalachian–Caledonian system illustrates this extremely well. The British Caledonides are an exceedingly complex assemblage of elongate disparate terranes, whose individual history can be related sensibly to plate boundary histories but whose present spatial distribution makes no plate boundary 'template' sense. This, perhaps, is the basis of David Howells complaint that plate tectonic templates, or 'Dewey-grams', have infected the regional tectonic literature so that geologists analysing and synthesizing a particular region have attempted to force disparate elements into 'Dewey-gram' cross sections. If regional geologists have used this methodology, this is indeed a disaster but 'Dewey-grams' are essentially plate boundary cross-sectional snapshots that apply strictly to actualistic present scenarios. Enforcing 'Dewey-grams' upon every orogenic cross section is doomed to failure, is not a practice that informed tectonicians follow, and is not a methodology ever used or encouraged by the writer in regional tectonic synthesis. However, there can be no question, as Alastair Robertson, Wes Gibbons, Donny Hutton and I emphasized during the meeting, that terrane thinking has changed the way in which we think about the Appalachian–Caledonian belt. The terrane complexity of the British Caledonides scarcely

189

could be understood without reference to the Newfoundland segment of the Appalachians in which strike–slip terrane motion along the critical western margin has been small. In Newfoundland, Ordovician arc collision and ophiolitic forearc obduction can be used, as shown by Dewey and Shackleton, to restore and explain the complex terranes of the British Caledonides.

In the Pan-African complex of the Afro-Arabian Shield, Harris, Gass and Hawkesworth argued that accretion of island arc terranes at a continental margin can result in rapid rates of crustal growth and that, in such cases, geochemical constraints provide strong evidence for allochthonous terranes in Precambrian belts. The Afro-Arabian Shield was produced by calc-alkaline magmatism between 900 and 600 Ma ago, where field evidence and trace element geochemistry suggest that Pan-African tectonics began as a series of intra-oceanic island arcs that were accreted together to form continental lithosphere over a period of 300 Ma. The great majority of Nd and Pb isotope ratios obtained for igneous rocks from the shield are indicative of a mantle magma source and estimated crustal growth rates are similar to those calculated from Phanerozoic oceanic terranes in the Canadian Cordillera. Such high rates in the Afro-Arabian Shield are consistent with the accretion of island arc terranes along a continental margin. Paul Hoffman warned, however, that rates of crust formation estimated from Nd model ages tend to be too fast because they give only the mean age of crust formation and, thereby, underestimate the duration of crust-forming episodes.

Paul Hoffman showed that cratonic North America is composed of a cluster of Archaean microcontinents, centred on the Canadian shield, and that juvenile early Proterozoic crust underlies the western and southern interior platforms. Formation of both the Archaean and Proterozoic crust is explicable by forearc accretion and magmatic arc progradation. Late Archaean and early Proterozoic accretionary systems differed through the involvement of older continental terranes that were narrower and wider, respectively, than the operative zones of collisional tectonism. The Archaean crust also differs from younger crust in having roots of buoyant refractory mantle tectosphere, possibly resulting from more prevalent physical underplating by depleted oceanic lithosphere. Hoffman believes that the buoyant tectosphere accounts for the biased exposure of Archaean crust in the shield and, conversely, the burial of Proterozoic crust beneath platform cover. Consequently, the isotopic ratios of detritus eroded from the craton overestimates the mean age of the continental crust, and sampling limited to the shield gives a false impression of predominant Archaean crust formation and Proterozoic epicontinental deposition.

Clearly, the terrane concept has been useful in emphasizing the role of large orogen-parallel strike–slip motions and the uselessness of the balanced-cross-section approach to orogens that evolve on the margins of large oceans. Most orogens consist of a collage of terranes of oceanic and continental origins that are assembled by the great natural complexities of plate boundary zone evolution. Terrane definition is only useful at the intermediate tectonic scale, otherwise their size ranges from the sand grain to the continent. Terrane boundaries are commonly intraorogenic transforms that excise and repeat earlier-assembled tectonic elements so that ocean margins orogens commonly assume a complicated braided pattern in which terrane boundaries with large motions commonly contain a wide range of exotic slivers, substantial mylonite zones with abundant constant shear sense indicators and palaeomagnetic evidence of large systematic rotations of upper crustal flakes and lack the characteristics of smaller displacements such as pull-apart basins with subjacent granite plutons, and transpressional zones. The margins of large oceans are characterized by regions such as the Andes from which terranes have been stripped and regions such as the U.S. Cordillera and Alaska to which terranes have been added. Smaller oceans lack this tectonic style and are typified by small-scale 'concertina' tectonics and/or windscreen-wiper tectonics whereby arc slivers leave one margin to collide with the other.

Perhaps the most energizing and illuminating aspect of the meeting, apart from the opportunity for a group of geologists to present and synthesize regional tectonic data from a wide variety of areas, was a heated and substantial wide-ranging epistomological discussion on the methodological relationship between hypothesis, model building and the collection of geological field data. Opinions varied from the view, expressed by John Sutton, that one should go into the field to collect data and to make geological maps untainted by hypothesis, to the other extreme, expounded by Celal Sengör, that one goes into the field strictly and only to test a hypothesis and that so-called objective data gathering is valueless. While leaning to the popperian end of the spectrum avowed by Sengör, I believe that much valuable, though limited, work has been done by 'objective' mapping untrammelled by hypothesis. However, as argued by Jack Soper, it must be true, surely, that the mindless acquisition of field data never leads to conclusions. Science advances by the complex iterative testing of ideas by data collection that shows these ideas to be wrong and by the injection of ideas that give us new ways of looking at the data and new kinds of data to collect. Lastly, the title of the Royal Society Discussion Meeting following this one may well have been used for this. Many geologists would regard terranes as Migrant Pests!

INDEX

192

INDEX

INDEX

196

INDEX